ISBN 978-1-331-89822-1
PIBN 10251351

1 MONTH OF
FREE
READING

at
www.ForgottenBooks.com

By purchasing this book you are eligible for one month membership to ForgottenBooks.com, giving you unlimited access to our entire collection of over 700,000 titles via our web site and mobile apps.

To claim your free month visit:
www.forgottenbooks.com/free251351

FUNGOID DISEASES OF AGRICULTURAL PLANTS

FUNGOID DISEASES OF GRICULTURAL PLANTS

BY

JAKOB ERIKSSON, Fil.Dr.

PROFESSOR AND DIRECTOR OF THE BOTANICAL DIVISION OF THE SWEDISH CENTRAL STATION
FOR AGRICULTURAL EXPERIMENTS, STOCKHOLM ;
MEMBER ROYAL ACADEMIES OF SCIENCE, STOCKHOLM, COPENHAGEN, AND ROME ;
CORRESPONDING MEMBER ROYAL SOCIETIES OF AGRICULTURE, VIENNA AND BRUNN ;
HON. MEMBER ROYAL HORTICULTURAL SOCIETY, LONDON, ETC.

117 ILLUSTRATIONS, OF WHICH 3 ARE COLOURED.

TRANSLATED FROM THE SWEDISH BY ANNA MOLANDER

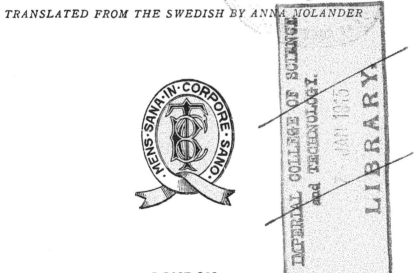

LONDON
BAILLIÈRE, TINDALL AND COX
8, HENRIETTA STREET, COVENT GARDEN

1912

PREFACE TO ENGLISH EDITION

THE increased facilities of intercourse between different countries which characterise modern civilization have proved an important factor in the diffusion of plant-diseases. The organisms associated with these diseases, which do untold damage by diminishing the world's supplies of food and fodder, pay scant respect to such human devices as tariff zones and political frontiers. But if the organisms in question be disquietingly cosmopolitan, the workers whose energies are devoted to the study of their life-histories, to the control of their activities, and to the mitigation of the evil consequences they induce, are also, fortunately, cosmopolitan. Nowhere is the solidarity of science more benignly manifested, and nothing connected with the amenities of human intercourse is more satisfactory than the neighbourliness with which these workers in every land place the results of their labours at the disposal of countries other than their own.

A more pleasing instance of this neighbourly feeling than the preparation of an edition, for the use of English cultivators, of Professor Eriksson's popular treatise on the diseases of plants due to fungi, could hardly be conceived. By no one will this digest of his wide knowledge, ripe judgment, and practical experience be more heartily welcomed than by his fellow-workers in the same field in the United Kingdom.

D. PRAIN.

Royal Gardens, Kew,
  April, 1912.

AUTHOR'S PREFACE

NOWADAYS we frequently hear the complaint that the diseases of our cultivated plants become year by year more numerous and more disastrous. It is alleged that new diseases are constantly turning up which have never before been seen or heard of. It is also said that parasitic fungi, which hitherto have proved quite harmless, have changed their nature, and become most destructive.

Is this really the case, or is it not? Some hold this opinion to be erroneous, and claim that the state is the same as before, but that greater attention and the strict investigation which now are given to this study bring the matter more before the public eye, and also result in the discoveries of hitherto unnoticed diseases.

This explanation can hardly be gainsaid. In most civilized countries there is now a diligent survey and a close inspection of the health of cultivated plants, and this attention must result in a detection of diseases that have been previously unnoticed. But, on the other hand, there are many cases that cannot be placed summarily in this category. And this refers especially to a great number of the fungoid diseases.

It is, after all, an incontestable fact that fresh fungoid diseases have recently appeared and are gaining a footing in various countries. How is this to be explained? Several causes work simultaneously to this end.

The inclination in our time for specialisation, even with regard to agriculture, inasmuch as only one or a few species are cultivated in large masses, helps to make the plants more susceptible. This mass-culture of varieties and sorts generates new characteristics in the individual plants, and some of these new forms will then be more susceptible to diseases and become a source of disease for the whole plantation.

But another important fact must be placed side by side with this. The law of evolution is known and accepted as a truth in the scientific world. Formerly it was thought that hundreds and even thousands of years were required for this evolution. But to-day we are certain that new forms can be brought forth through impulsive new formation—" mutation "—and these forms can vary from the parent stock in one or more characteristics. Why could not this natural law be applied also to the extensive dominion of the parasitic fungi? And why could not these new forms of fungi possess qualities that would enable them to avail themselves of the nourishment offered through the new races of host-plants brought about by the mass-cultivation? This is quite possible, and recent experiments and observations tend to prove this supposition.

Beside these two purely scientific explanations, we have the practical one of the easy communication of our days, by means of which contagious matter can be readily spread from one district, country or continent, to another.

The combined effects of these circumstances render the research and the prevention of fungoid diseases a matter of international importance.

In this work is given a general review of all important fungoid diseases that attack agricultural plants in the countries of Northern Europe, and also of available means for combating with or preventing the said diseases. More than 200 different diseases are treated.

In this work are included not only diseases which have established themselves in the above-mentioned countries, but also those that have been noticed in other countries. With the quick communications of our days and the unceasing transactions between different countries, both in export and import, it does not take long before diseases spread to places where they previously have been unknown.

In order to render it easier to detect different diseases, illustrations are attached. We have in each case named the authors from whose works the illustrations have been borrowed. Our own are marked: " The Author."

As the primary object of this book is to be a practical guide for planters, to enable them to recognize, prevent, and battle with the diseases, historical and literary facts generally have been omitted.

The celebrated phytopathologist, Mr. George Massee, of the Royal Gardens, Kew, has done me the honour of reading over the English manuscript with a view to checking the technical terms throughout. For this act of kindness I wish to express to him my most hearty thanks.

THE AUTHOR.

EXPERIMENTALFÁLTET, STOCKHOLM.

CONTENTS

LIST OF ILLUSTRATIONS

INTRODUCTION

The Structure and Life of Fungi.

FUNGI form a large group of their own in the vegetable world. In appearance they differ—like algæ and lichens—from the higher, more developed plants in that they do not show any difference between root, stem, and leaves. The whole body of a fungus consists simply of a *thallus* of varying shape and structure. In their inner nature and their mode of living the fungi differ from all other groups of plants by the absence of chlorophyl, the lack of which renders it impossible for them to directly absorb and assimilate inorganic things, such as carbonic acid, water, nitric acid, and ammonia, and with them form organic compositions. They are destined to obtain their nourishment, ready prepared, from other organic bodies, either living or dead. Fungi which obtain their food from living animals or plants are called *parasitic* fungi; those that live upon remains of either animals or plants are called *saprophytic* fungi. The difference between them is, however, not very distinct. There are some fungi which appear sometimes under one of these types, sometimes under the other.

That part of the body of the fungus which accumulates and distributes the nourishing matter is called the *vegetative* system. This is, as a rule, a spawn (*mycelium*) consisting of articulated and ramified filamentous tubes (*Mycomycetes*). In the simplest forms of fungi this spawn is reduced to a single filamentous tube, usually without articulation (*Phycomycetes*), or it is replaced by an irregularly shaped mucous body—a *plasmodium* (*Myxomycetes*)—or it may be missing altogether (*Schizomycetes*).

xiii

When the spawn has reached a certain degree of development the fructifying system of the fungus appears, and produces breeding organs—*germ cells* (*Conidia*), or *spores*. The conidia germinate at once. The spores are various in structure and different in nature. With regard to their shape, they vary, being ball- or egg-shaped, club-like, thread-formed, or semicircular. In structure they are one-celled up to many-celled. On the surface they are either smooth or shrivelled, or have warts, hair, or bristle. Some of them germinate as soon as they arrive at maturity (*summer spores*); others require a certain resting period before they grow (*resting*, or *winter spores*).

Synopsis of the Different Groups of Fungi.

In this work only such parasitic fungi will be noticed that obtain their nourishment from agricultural plants.

These fungi can be divided into the following groups:

I. SPURIOUS FUNGI, which have no filamentous tubes.

    1. **Schizomycetes.**

    2. **Myxomycetes.**

II. GENUINE FUNGI, which have a spawn of filamentous tubes.

(i.) Simpler fungi (*Phycomycetes*), the spawn of which consists of one single, mostly inarticulate, but often much ramified, filamentous tube.

    3. **Chytridiaceæ.**

    4. **Peronosporaceæ.**

(ii.) Higher fungi (*Mycomycetes*), whose spawn consists of numerous filamentous articulated tubes.

    (*a*) Basidiomycetes.

    5. **Ustilaginaceæ.**

    .6 **Uredinaceæ.**

    7. **Clavariaceæ.**

(*b*) Ascus fungi (*Ascomycetes*).

8. **Erysiphaceæ.**
9. **Perisporiaceæ.**
10. **Sphæriaceæ.**
11. **Nectriaceæ.**
12. **Dothideaceæ.**
13. **Pezizaceæ.**
14. **Helvellaceæ**

(*c*) Imperfect Fungi.

15. **Sphæropsideæ.**
16. **Hyphomycetes.**

    (α) *Melanconieæ.*
    (β) *Tuberculariaceæ*
    (γ) *Mucedineæ.*
    (δ) *Dematieæ.*
    (ε) *Rhizoctonieæ.*

FUNGOID DISEASES

OF

AGRICULTURAL PLANTS

CHAPTER I

BACTERIA—SCHIZOMYCETES

BACTERIA are the smallest and morphologically the simplest amongst organized beings. When aggregated in millions they appear to the naked eye only as the finest sediment, whether in liquid putrefaction or in the tissues of diseased animals or plants. Each individual consists of only a single cell. It, however, frequently happens that several cells are in process of forming colonies, and it is an accumulation of these colonies or congeries that constitutes the sediment referred to. They can be defined only under high powers of the microscope. The shape of bacteria is variable, as seen in the four forms on p. 2—viz., the spherical, the rod-like, the spiral, and the flagellate. It is particularly the spherical and the rod-like that are active in the origin of plant diseases.

Bacteria increase, as a rule, by fission. The two cells may either begin an independent existence, if separated, or otherwise together develop thread-like, tabular, or cube-like colonies. Bacteria are frequently provided with one or more vibratile organs, which serve as a locomotive power. However, there has been noticed in the case of many bacteria an increase, not only by fission, but also by spores. The spores are thick-walled, and originate inside the cells of the bacteria. This development of spores appears especially to take place when the nourishment is scanty, and as a sequel the vegetative process abates. These spores might be styled " resting spores." They possess great resistance against outside influence.

It is a long-known fact that bacteria are the agents not only in putrefaction and fermentation, but also in infectious and contagious diseases, both in man and beast. But that they are also the cause of plant diseases is known from the observations, experiments, and researches of the last few decades. And every year our knowledge is increased by the discovery of some additional disease of the same kind, known under the name of *Bacteriosis*. This new science is undecided as to whether bacteria are actual parasites which originate the prevailing sickness in the plant, or if they possibly may have settled on the previously diseased and decaying vegetation. In the following pages an account will be given of several of the most important and most thoroughly investigated forms of bacteriosis.

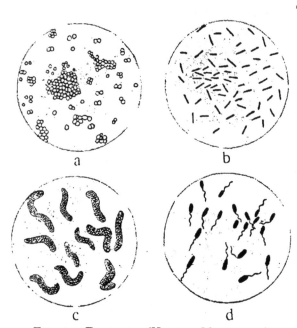

FIG. 1.—BACTERIA (HIGHLY MAGNIFIED). (FROM W. MIGULA AND E. WARMING.)
a, Spherical ; *b*, rod-like, or bacilliform; *c*, spiral ; *d*, flagellate.

Potato-Tuber Bacteriosis. (*Bacillus amylobacter* and *B. solaniperda*.)

This disease appears in the autumn at the time when potatoes are taken up from the ground. In the first stage it appears as brown spots within the tubers, either as several small spots scattered inside the tuber, or as one large spot in the very centre. If the soil is clayey, the season rainy, and the potatoes are stowed away in consequence in a damp condition, then the rot will spread all over the tuber. Not only the walls of the cells, but also the starch granules dissolve, until the whole inside of the tuber consists of a fetid, thick matter, more or less covered by the yet unaltered skin. The potato is now said to be " wet-rotten."

Should it, on the contrary, happen that the weather is very dry and the soil of a sandy character, the drainage therefore being good, then the progress of putrefaction will cease. The tuber dries and gets hard, and inside there develop many little vacua, lined with a white layer of starchy material. In this case the potato is said to be "dry-rotten." The starch in such a

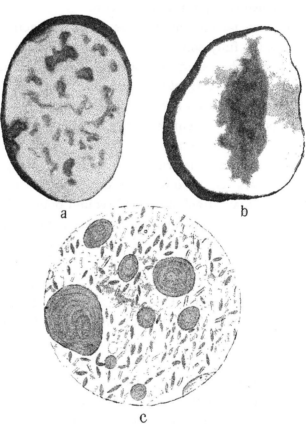

FIG. 2.—POTATO - TUBER BACTERIOSIS (*c*, HIGHLY MAGNIFIED). (*a* AND *b*, FROM E. ROSTRUP; *c*, FROM C. WEHMER.)

a and *b*, Sections of diseased potato-tubers ; *c*, part of the slimy matter, with two different bacteria : one of them (*Bacillus amylobacter*) pointed towards the ends, the other (*B. solaniperda*) with straight-cut ends.

potato can be used for technical purposes, such as dextrine and other similar products. It was thought for some time that wet-rot, as well as dry-rot, was caused by the potato mould (*Phytophthora infestans*), and it was considered that the brown spots produced by this fungus on the surface of mature potatoes were a previous stage of both forms of rot. Finally, an agreement has been arrived at that this is not the case, but that wet-rot, as well as dry-rot, is developed solely through bacteria.

But so far the question as to which bacteria produce these two rots is not fully solved. Scientists have arrived at different results. It is now, however, generally believed to be the *Bacillus amylobacter* (also named *Amylobacter navicula*, or *Clostridium butyricum*), and *B. solaniperda*, both bacilliform bacteria.

A point of contention has been whether the bacteria can of themselves develop disease in a perfectly sound potato, or if they can only complete a destruction introduced through some other agent. It appears as if both these can take place. In some instances bacteria may attack a perfectly sound potato, if it be injured in some way, so as to give them an entrance; but especially with regard to *B. solaniperda* the potato can be infected without any visible defect on the surface.

The power of resistance against these various attacks is variable with different sorts of potatoes. The most susceptible are the early food potatoes, while the late fodder and factory potatoes have the greatest resistance.

PROTECTIVE MEASURES.—(1) Never use anything but perfectly sound seed. (2) Select for larger crops only such varieties as have been proved in the district most capable of resistance. (3) Grow potatoes only in such soil as is high placed, well drained, and mixed with sandy material. (4) Take care that the soil which is cultivated be rich in potassium and phosphate, as an abundance of these has proved to increase the resistance against rot, while, on the other hand, strong lime manuring produces the opposite effect. (5) The potatoes must needs be quite dry when put away for winter supply, and the place where they are laid up should be dry and well ventilated.

Potato Ring-Bacteriosis. *(Bacillus Solanacearum.)*

Occasionally it happens that a few weeks after the potatoes have been planted, several of the seeds do not sprout. If you then dig at the vacant place, you will find the seed-tuber perfectly sound in appearance, but the sprouts are fading away, and also the abnormally richly developed roots. Other plants reach beyond the ground, but soon acquire a sickly and dwarfed appearance, with a glassy stem and petty leaves. Finally, the growth of the plants is suspended altogether. Several other plants continue to grow in seemingly good state until summer-time, when their hearts become transparently brown-spotted, and wither away prematurely. These last-mentioned specimens yield almost a satisfactory crop.

This disease, also named " Potato Wilt-Bacillose," is generated by bacteria, especially *Bacillus Solanacearum*, which follow the fibro-vascular system of the whole plant, and originate black spots. If the case be only a slight one, there appears on the section a sickly-looking ring towards the surface of the potato. In more serious cases the whole inside turns dark. Frequently no sign of disturbance can be noticed on the skin.

It is particularly through slightly infected tubers that the disease is maintained year by year. If a tuber of such quality be planted, the plant itself will become more or less diseased. The disease will spread to sound potatoes through some defect on their surface, thus making an opening for the bacteria.

This disease has sometimes proved very disastrous. In the year 1905 in one district of Germany 60 to 70 per cent. of the crop was destroyed—in fact, there were fields so badly attacked that it was deemed futile to attempt to gather the crop.

PROTECTIVE MEASURES.—(1) Use for seed, only strictly sound potatoes. An investigation should be made in such a manner, that the potatoes be cleft and placed in the air with the cloven sides upwards. If in an hour or two there appear black pricks and spots on the surfaces, it indicates the existence of the disease, and these specimens should be rejected as seed-tubers. (2) If possible, whole and not divided tubers should be used for planting. If, however, it is desirable to

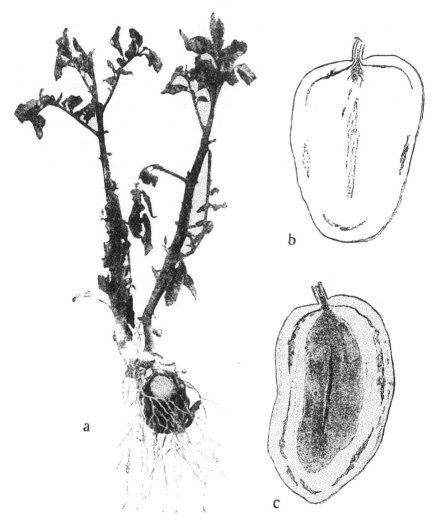

FIG. 3.—POTATO RING-BACTERIOSIS—*Bacillus Solanacearum.*
(FROM O. APPEL.)

a, Plant spired from a ring-diseased potato ; *b*, a slightly, and *c*, a seriously, ring-diseased potato in section.

use divided tubers, then the division should take place a couple of days previously, in order that there may accumulate on the cleft sides a cork layer that will prevent the entrance of the bacteria.

(3) Examine during the summer the potato-fields, and destroy thoroughly both stalks and tubers of all prematurely withered specimens as soon as they are suspected of propagating the disease.

FIG. 4.—POTATO-STALK BACTERIOSIS—*Bacillus melanogenes* AND OTHERS.
(*a*, FROM O. APPEL ; *b*, THE AUTHOR.)
a, A whole plant, diseased ; *b*, the lower part of a diseased stalk.

Potato-Stalk Bacteriosis. (*Bacillus melanogenes*, and others.)

This disease, also called "Black Stalk-rot," becomes conspicuous in the potato-fields early in the summer through the way in which the leaves turn yellow and cease growing, while the heart-leaves turn slightly upwards. Simultaneously that part of the stalk which is in the ground gets black and withers away, and shrinks narrower than the upper stalk. Often only a single sprout of the plant may

thus be affected.   If the weather is wet the black part of the stem
will appear to be slimy, and bacteria are to be found in the vascular
system of the stalk proper.   Towards the end of June plants of this
description appear amongst the healthy ones.   By this time no
tubers have begun to form, and within a week the plant dies.
With regard to those plants which become affected later in the
season, and in consequence have formed tubers, the infection will
shortly reach them.   The inside of these tubers turns black and

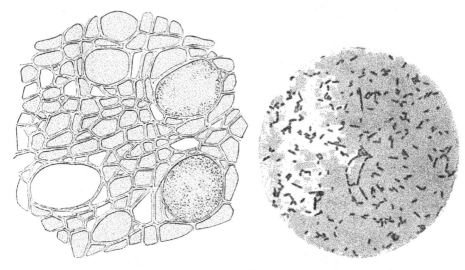

Fig. 5.—Potato-Stalk Bacteriosis.   Fig. 6.—Bacillus phytophthorus
(From O. Appel.)   (Highly Magnified).   (From
Vascular tissue, with two vessels leading   O. Appel.)
    the bacteria upwards from the
    bottom part of a diseased stalk. ·

begins to rot.   Slightly affected tubers, used afterwards as seed,
develop sickly offspring.

From one country and another reports are obtained giving
different opinions as to the bacteria that originate the complaint:
Bacillus caulivorus (France), *B. atrosepticus* (Holland), *B. phytophthorus*
(Germany), *B. solanisaprus* (Canada), and *B. melanogenes* (Ireland).

This disease was first noticed and investigated in France, where it
was called "Gangrène de la tige."   This took place about the year
1890.   In Germany it goes by the name of "Schwarzbeinigkeit."

Both these names could appropriately be translated into English as " Black Leg." In Sweden it appeared in 1908 and 1909, and spread over numerous fields in various provinces.

PROTECTIVE MEASURES.—(1) Use sound seed. (2) Use in preference undivided potatoes for seed. (3) Do not cultivate potatoes in soil where during previous years sickly plants have grown. (4) Avoid nitrogenous fertilizing. (5) Examine the potato-fields frequently during the summer, and pull up and destroy at once all sickly plants, thus preventing the ailment from spreading to sound plants in the same field.

Potato Brown-Bacteriosis.

This name might apply to another sort of stalk-bacteriosis, which appears later in the summer, and for the first time was noticed and described in France during 1901. It was there called " Brunissure," and was observed in England somewhat later. The leaves turn yellow prematurely. The stalks become thin and wither from the lower part. The ring of vessels develops yellowish-brown spots, with suppurating yellow slime both in stalk and tubers. The disease is originated through *B. solanincola.*

Cabbage Brown-Bacteriosis. *(Pseudomonas campestris.)*

This disease, usually called "Black Cabbage-rot," attacks different sorts of cabbage, especially white cabbage and cauliflower, but also Swedish turnip, turnip, rape, and others. In the white cabbage it usually appears first at the edge of the leaves as a brownish-black veiny mesh, and passes from there into the leaf-stalks and stalk proper. In the Swedish turnip and the turnip it reaches the root.

Bacteria evidently gain an entrance either through the water-pores at the edge of the leaves, or else through some wound somewhere on the surface of the plant. They accumulate in the vessel-strings, that in a cross-cut section appear as brownish-black pricks, distinguishable to the unaided vision.

The bacteria continue their work of destruction with the crop stored up for the winter. Even the most solid and perfect cabbage tops are then affected, until the whole heart is transformed into a soft, spongy, black-spotted lump.

In turnips the disease manifests itself in somewhat different forms. It sometimes appears in the shape of dark stripes arranged radially; at other times, again, these stripes are longitudinal, in both cases inside the turnip, while the surface appears

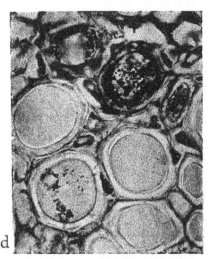

FIG. 7. — BROWN-BACTERIOSIS IN WHITE CABBAGE — *Pseudomonas campestris* (*d*, MAGNIFIED). (*a, b, c*, FROM E. ROSTRUP; *d*, FROM E. F. SMITH.)

a, A cabbage-leaf, in which a part of the veiny mesh is blackened by bacteria; *b*, section of a leaf-stalk, with vessel-strings black-spotted by bacteria; *c*, part of a cabbage-stalk, with black-spotted scars after the diseased leaves have dropped off; *d*, section of a vessel-string filled with bacteria.

comparatively clean. Finally, the whole inside turns into a slimy, fetid mass.

With regard to their shape, the diseased turnips differ from the sound only in this respect, that they do not acquire the ordinary broad, spherical or flat shape, but grow long and thin, almost like

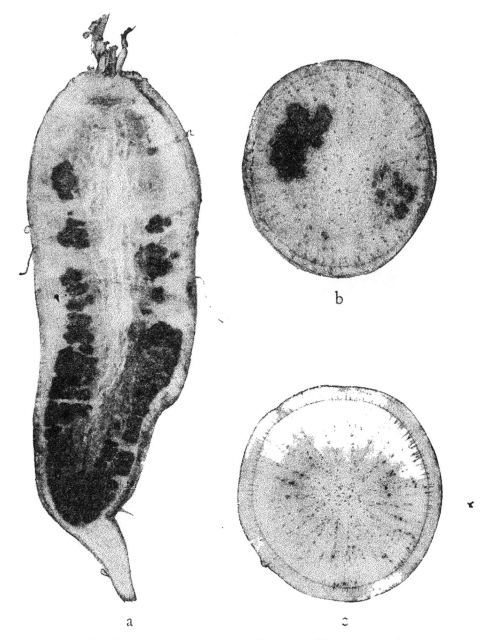

FIG. 8.—BROWN-BACTERIOSIS IN TURNIP. (THE AUTHOR.)

a, Section of turnip lengthwise, with two longitudinal diseased strings or stripes ; *b,* a similar turnip cut across the root ; *c,* a cross-cut turnip (from another locality) with only radial diseased strings.

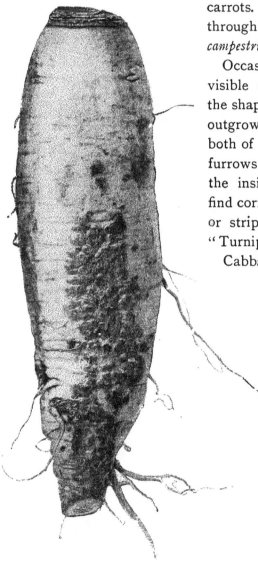

carrots. The disease is brought on through a bacillus called *Pseudomonas campestris*.

Occasionally the disease appears visible on the outside of the root in the shape of an abundance of wart-like outgrowths of the size of a pea along both of the two longitudinal depressed furrows on the surface of the root. In the inside of the root you will then find corresponding longitudinal strings or stripes. This form can be styled "Turnip Wart-Bacteriosis."

Cabbage Bacteriosis was first investigated and described in the United States of America in the year 1895. At the present time it is said to be very prevalent in that country, usually under the name "brown-rot," or "black-rot." In Europe it has turned up many times during recent years, as in Holland; France, "Nervation noire"; Switzerland; Germany, "Schwarze Fäulniss"; and Denmark. Quite lately it has been noticed in Sweden, in the year 1906 on white cabbage and turnip, and 1907 on turnip.

FIG. 9.—TURNIP WART-BACTERIOSIS.
(THE AUTHOR.)

PROTECTIVE MEASURES.—(1) Pinch off and destroy all leaves that are attacked, and if the disease has already reached the inside of the plant it should be destroyed altogether. (2) Do not let diseased fragments of plants lie about the field during the winter. (3) Do not gather seed from infected fields. (4) Seed, which you suspect, should before sowing be soaked in a mixture of sublimate (1 in 1,000) for fifteen minutes, or in a mixture of formalin (1 in 200) for twenty minutes. (5) For two to three years no cabbage should be cultivated in a field that has been infected. (6) Keep the field cleared of charlock and other Cruciferæ, as these also might have been infected, thus retaining the bacteria in the field.

Besides these, there have been reported the following new forms of Bacteriosis of cabbage :

White-Rot of Turnip, noticed in England in 1900, with leaves turning yellow and dropping off—first the outside leaves and then the younger; the diseased parts of the roots are greyish-white; finally the whole root becomes putrid and offensive ; the disease originates through *Pseudomonas destructans.*

Bacteriosis of White Cabbage, noticed in Germany in 1902, and beginning in the younger stalks and in the central nerves of the leaves, the diseased parts dissolving into an offensive pulp. The disease is provoked through a species of *Pseudomonas*, which has lodged between the cells.

Bacteriosis of Cauliflower and other sorts of cabbage, caused by *B. oleraceæ,* and discovered in Canada in 1901, and also by *B. brassicævorus,* noticed in France in 1905.

Beet Mucous-Bacteriosis.

Of this there are several sorts. One of them—" Beet gummosis," " Bacteriose gummosis," or " Gummose bacillaire "—indicates its existence in such manner that the vessel-strings in the root first turn reddish-brown, then black. When cut longitudinally, there appear dark stripes, running lengthwise. If the section be made transversely, dark rings appear. Finally the whole interior dissolves into a glutinous or molasses-like slime. This disease has been noticed in many places in Europe, especially in Austria, Germany, Belgium, and Denmark. It is caused through several different forms of bacteria : *B. Betæ*, *B. lacerans*, *B. Bussei*, and others.

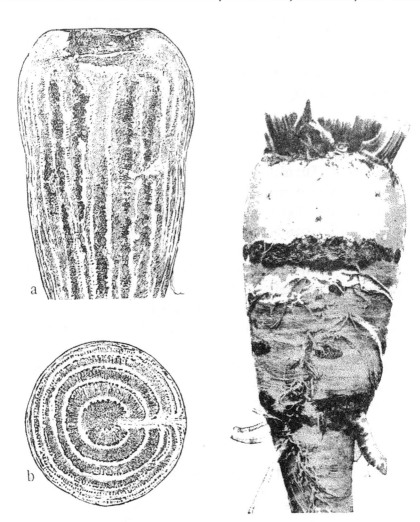

FIG. 10.—BEET MUCOUS-BACTERIOSIS —*Bacillus Betæ* AND OTHERS. (FROM P. SORAUER.)

a, Longitudinal section ; *b*, transverse section.

FIG. 11.—BEET WART-BACTERIOSIS —*Bacterium scabiegenum*. (FROM F. C. VON FABER.)

Beet Wart-Bacteriosis.

This "Pustelschorf" appears as small black warts on the surface of the root, sometimes with a crater-like concavity in the centre of the wart. In sugar-beets the warts often together form vertical swellings. In fodder-beets the warts appear mostly on the lower part of the root. The disease rarely penetrates deeper into the substance. In many cases the wound-spots heal and the warts drop off. In no case has there been found any general putrefaction. The disease has been noticed in Northern and Central Germany, and in the United States of America. It is caused by *Bacterium scabiegenum*.

Beet Yellow-Disease.

The "Yellowing of beet leaves" or "Jaunisse" manifests itself on the leaves, first the outside leaves in the bunch, then the inner. The leaf-stalk shows alternately light yellow and dark green spots. It assumes a kind of mosaic appearance, best seen if the leaf is held towards the daylight. From the leaf-stalk the disease goes into the plant-stalk, which turns into a glassy transparency, first in the upper part, thence lower down.

a   b

Fig. 12.—BEET YELLOW-DISEASE —*Bacillus tabificans*. (THE AUTHOR.)

a, Early stage, with mosaic-chequered leaf; *b*, later form, with a withered leaf and the leaf-stalk transparently glassy.

The leaf feels sticky and soon dies. If the beets bloom, the same sickly manifestations appear in the flowers. Should the attack be severe, it will arrest the growth of the root. The disease was observed for the first time in the North of France in 1896. In South Sweden its existence was proved in the beginning of August, 1909. It is brought on by *B. tabificans.*

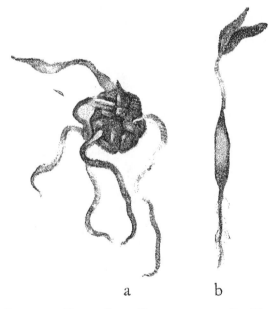

a   b

FIG. 13.—BEET GERM-BACTERIOSIS — *Bacillus mycoides.* (FROM G. LINHART.)
a, Seed-bundle, with protruding diseased sprouts;
b, a diseased young germinating plant.

Beet Germ-Bac-teriosis.

This attacks with more or less disastrous results the tiny sprouts when they protrude from the seed. In severe cases these are totally destroyed. The disease is especially studied and mentioned in Hungary. It is principally caused by *B. mycoides.*

Californian Beet-Pest.

The " Beet-pest " has been noticed in California since 1899. The leaves become small, yellowish-brown, finally black ; the root dwarfed, not larger than a radish, with plenty of root-strings. The interior of the root shows dark concentric rings. From the vessels penetrates a dark juice, which turns black in the air.

PROTECTIVE MEASURES.—(1) Pull up and destroy plants immediately they get diseased. (2) Do not let diseased remnants lie in the field over the winter. (3) Be careful that soil from an infected field is not carried by means of vehicles, implements, dray-animals and other things to the field that is intended for next year's beet-crop. (4) Do not allow the remains of diseased beets to get mixed with the manure. (5) Do not take seed from beets grown in soil that was infected. (6) In a field that produced diseased beets, no beets should be planted for at least three years.

Amongst other diseases of agricultural plants considered to originate from bacteria may be mentioned :

Seed Bacteriosis.—The diseased seeds, of a rosy hue, develop poorly, become shrivelled and sometimes even hollow. The disease has, since 1878, been occasionally noticed on wheat in France, and also in Denmark. It is caused by *Micrococcus Tritici.* In Sweden, a similar disease has been noticed on two-row barley.

Lupin Bacteriosis.—Young plants develop first yellow and then brown spots on the leaves, and shortly wither. It was noticed in 1899 in Hungary, and is caused by *B. elegans.*

Carrot Bacteriosis, or "Soft-rot of Carrot."—The carrots are attacked by a soft-rot while stored up for the winter, usually starting at the crown, and quickly proceeding downwards. The affected part becomes mellow and brown. This disease acquired a malignant type in 1897 and 1898 in Vermont, U.S.A. It has been proved to spread to many other plants, as turnip, rutabaga, radish, salsify, parsnip, onion, celery, tomato, and others. It is caused by *B. caratovorus.*

Tobacco Bacteriosis.—Of this there are several kinds. One of them, "Chancre bactérien," forms on the stem and on the principal nerves of the leaves in long, groove-like, dark spots, and may cause the death of the plant. The disease is caused by *B. æruginosus,* and is widely distributed in France.—Another disease, "Maladie des taches blanches," forms on the leaves small, irregularly shaped, at first pale green, then white and dried spots. It is caused by *B. maculicola,* and, like the previous one, is widely spread in France.—A third one, "Granville tobacco-wilt," is prevalent in North Carolina, Florida, Georgia, U.S.A. The leaves become dry, bend down, and die. The layer of wood in the stem turns black and dies. The disease is thought to be caused by *B. Solanacearum* (see previous account of Potato Ring-Bacteriosis).

CHAPTER II

SLIME MOULDS—MYXOMYCETES

THESE fungi, which are of a very low order, do not consist of real cells or filamentous tubes, but have only a naked mucous body (*plasmodium*), lacking in clearly defined outlines. When maturity is reached, the plasmodium is transformed into a large number of spherical spores. When a spore germinates, there develops a roaming body (*myxamœba*), which is continually changing its shape and progresses by creeping movements. The mucous bodies take their sustenance from the substance—either dead or alive—upon which they grow, and often many of them collect into one large mucous mass.

Most of the slime moulds live upon a dead substance, such as decayed wood, rotten stumps of trees, withered leaves, and so forth; and then it frequently happens that they further extend to living plants or to parts of them. Thus they can do damage by depriving them of light and air. The saprophytically existing forms possess as a rule minute fruit bodies (*endospores*), with a wall formed from dried mucus, and an inside mesh of fine threads.

A few of the slime moulds appear as parasites on living plants. In their case the common covering round the spores is missing. The spores are contained in the cells of the plant that support them.

Cabbage Club-Root. (*Plasmodiophora Brassicæ.*)

The history of this disease can be traced back to the year 1736, when it was spoken of in England; but not until 1820 was it considered as a harmful destroyer from an economical point of view. After that time it is described as bad in many lands.

This disease thrives not only on the different varieties of *Brassica oleracea* in our kitchen gardens—viz., white cabbage, cauliflower, nettle-cabbage, brussels-sprouts, etc., but also on other cabbage-like plants, such as Swedish turnip (*B. Napus*) and turnip (*B. Rapa*).

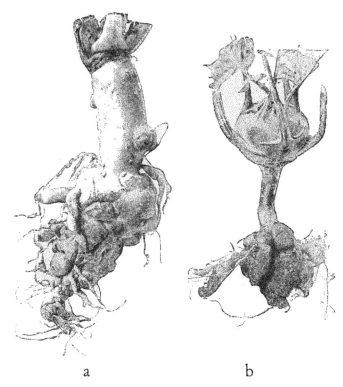

a   b

FIG. 14.—CLUB-ROOT DISEASE—*Plasmodiophora Brassicæ.* (THE AUTHOR.)
a, In a turnip ; *b*, kohlrabi.

Further, it has been noticed on numerous other plants of the Cruciferæ family, cultivated as well as wild—viz., *Capsella Bursa pastoris, Cheiranthus Cheiri, Erysimum cheiranthoides, Hesperis matronalis, Iberis umbellata, Matthiola incana, Raphanus Raphanistrum, R. sativus, Sinapis alba, S. arvensis, Thlaspi arvense,* and others ; and quite lately in France, although only once, in water-melon, celery, and sorrel.

The disease manifests itself in the cabbage and turnip fields during the summer in such a way that sundry plants cease growing and die away. If the roots of these plants are examined, they will be found covered with nodose swellings much varied in size and shape. The

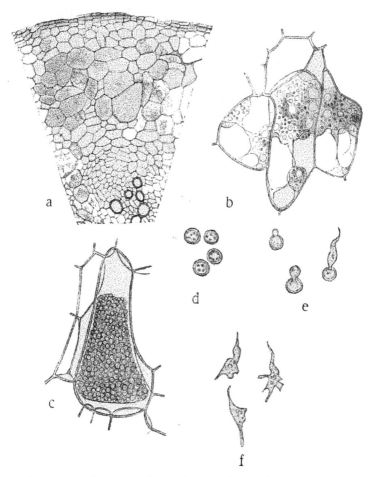

FIG. 15.—CABBAGE CLUB-ROOT. (FROM M. WORONIN.)

a, Section of a diseased root ; *b*, cells from the root-tumours, with mucous bodies ; *c*, cell, with spores ; *d*, spores ; *e*, germinating spores ; *f*, swarming bodies.

root appears thus very much distorted. Until the end of the summer these tumours keep white and hard ; late in autumn they rot, and become a dark, fetid mass.

If one of these tumours be examined under a microscope during summer-time, it will be seen that several of the cells are more or less filled with a colourless mass of fine granular particles. If high magnifying is applied, it will be seen that these granules consist of mucous bodies enclosed in the interior of the cells. These bodies belong to the fungus of the club-root disease (*Plasmodiophora Brassicæ*).

They pass through several stages of development. At first they are separated from each other, but later on float together into a common plasmodium, which more or less fills up the whole cell. Finally, the whole plasmodium resolves itself into numerous spherical thick-walled spores. When the diseased roots decay in autumn these spores become free in the surrounding soil. They remain unaltered during the winter, but in the spring they retain the power of germination. From the spore the mucous substance exudes, forming a movable swarming body of a protean nature and constantly changing shape. If such a swarming body strikes upon a young root-string of a susceptible plant, it will force its way into it. The root is then infected. Soil which has been infected retains the infection three years for certain—most likely five years, and possibly longer.

This disease will be carried from a sickly to a sound field by means of agricultural implements, the hoofs of dray animals, and the footgear of men. It is not proved that any considerable distribution is caused through manure from live stock having fed on diseased roots, except in those cases when remains of the fodder, which possibly have been left in the manger, get into the manure.

In certain cases it seems as if the disease is propagated with the seed. Thus it happened in 1903, on an estate in Sweden, that the disease appeared only with turnips raised from fresh seed, while plants of the old stock were sound all over the field. The disease varies in its ravages in different years and soils. Fields on a low level promote it; chalky soil resists it.

PROTECTIVE MEASURES.—(1) If it be the first year that the disease appears in the place, then pull up and remove early in the summer, before the root has begun to rot, every one of the sickly plants and every string-root. (2) If possible, do not cultivate for a period of four to six years any kind of cabbage in a field that previously has produced a diseased crop, but plant instead other vegetables, as potatoes, peas, clover, straw-seed, etc., and take care that the soil meanwhile is kept perfectly clear from weeds of the genera *Sinapis, Thlaspi, Capsella, Erysimum,* and others, on which the fungus can find its sustenance. (3) If it should be imperative within a year or two to again plant cabbage in sickly soil, then mix with the earth a year and a half, or at least half a year previous to the planting, an abundant supply of finely ground chalk, and not less than 2 tons

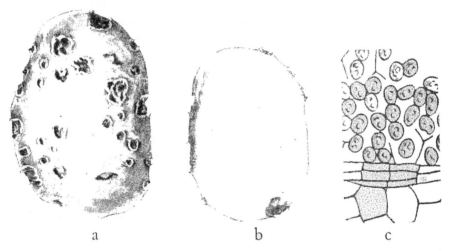

a   b   c

Fig. 16.—Potato Corky-Scab—*Spongospora scabies.* (From T. Johnson.)
a, A whole potato; *b,* potato (cross-section); *c,* fungi-carrying cells close below the potato skin.

slaked lime, or 4 tons carbonate of lime per acre. Kainite has also been used with success; about 15 cwt. per acre, sprinkled around in the spring. (4) Be careful to prevent the spread of disease from infected to sound earth through implements, hoofs of animals, and footgear of the agricultural workers. (5) Do not use any earth in hotbeds for cabbage that is not perfectly free from suspicion of infection, and cleanse carefully all the frames used for the purpose. Examine also every plant before planting out, lest there should be some indication of tumours on the root. (6) If slightly diseased roots are to be used as fodder, then care should be taken that no remains of these get mixed with the manure; the best way would be to boil the roots before feeding the stock with them. (7) If it be preferable not to use any sickly roots for fodder, then the diseased roots should be absolutely got rid of by burning or otherwise, but by no means mixed with heaps of dung or compost.

Potato Corky-Scab. *(Spongospora scabies.)*

This disease manifests itself in the shape of small concave spots with turned-up edges all over the skin of the tuber. If the soil be comparatively dry, a healing layer of cork forms beneath the sickly spots, and the disease stops. But in very moist earth no such healing process takes place. The fungus penetrates deeper and deeper into the potato, and the cavities continue to increase in size.

The cells in the sickly parts of the tuber contain the fungus which provokes the disease *(Spongospora scabies)* in its different stages of development. At first it appears as a number of spherical plasma bodies, then as one single plasmodium, in shape somewhat like a bunch of grapes; and, finally, as a mass of number-less spores.

FIG. 17.— SPUMARIA ALBA ON GRASS. (THE AUTHOR.)

The disease is reported from Norway, Ireland, and England. Especially in Ireland is it said to be widely spread, and is there considered as seriously destructive. It has also appeared in isolated localities in Germany, and was noticed in the autumn of 1908 in the western part of Sweden.[1]

PROTECTIVE MEASURES.—(1) Do not use diseased potatoes for planting. (2) If this cannot be avoided, then the diseased potatoes, as

---

[1] By the name of scab several other diseases of potatoes are known, although they have no connection with the one just described. These are caused by different conditions of the soil, through manure, or by artificial fertilizing.

well as those that have a sound appearance, but are gathered from aninfected field, should be steeped before planting in a 2 per cent. Bordeaux mixture or sulphate of copper. The steeping should last about twelve to eighteen hours. (3) For at least three years no potatoes should be planted in an infected field.

As indirectly injurious, slime moulds on grass lawns may be mentioned:

Physarum cinereum appeared on pastures in Sweden in 1905. The straws were covered by a mucous mass, and soon acquired a greyish-white hue, which finally turned black. There appeared a kind of black dust, together with greyish-white remains of cracked skin. The fungus arrested the development of the straw.

Spumaria alba appeared in the same place during the following years, also on grasses, and took the shape of large, snowy-white or grey-black mucous masses, especially on the root-stumps left after the hay had been cut.

CHAPTER III

CHYTRIDIACEÆ

THE fungi of this class are the most inferior of the fungi that, as a rule, are provided with *mycelium*. The mycelium, when such exists, is, at the beginning at any rate, without partition walls. Every new individual fungus originates from a spore, which settles upon one of the cells of the host; whereupon it penetrates into the cell, and there completes its development. Most of these fungi are aquatic, and exist as parasites upon algæ. Only a few settle as parasites on plants of a higher order.

Potato Black-Scab. (*Chrysophlyctis endobiotica.*)

This disease, also called "Wart Disease," "Cauliflower Disease," "Canker Fungus," and "Potato Rosette," appears as warty, uneven growths on the new tubers, which develop when the potato is flowering. Sometimes these warts are also on the sprouts that develop the new tubers, or else on other parts of the stem and on the roots. As a rule these growths reach the size of a walnut, or even larger. In very serious cases the whole thing finally looks like a huge, irregular mass of tumours. These growths, which at first are white, later on acquire a dark brown or black colour. If the sick potatoes are stored up in a damp state, then they will soon all rot away.

This disease is caused by a fungus—usually named *Chrysophlyctis endobiotica*—which lives within the potato cells, especially in the parts nearest the skin. In every diseased cell is a plasmodium, which by degrees is transformed either into one to three spore-cases of yellowish-brown colour, each containing a vast number of vibratile spores, that are agents for the spread of the fungus during the time of vegetation, or else into one single thick-walled resting spore, that retains the fungus alive until another year.

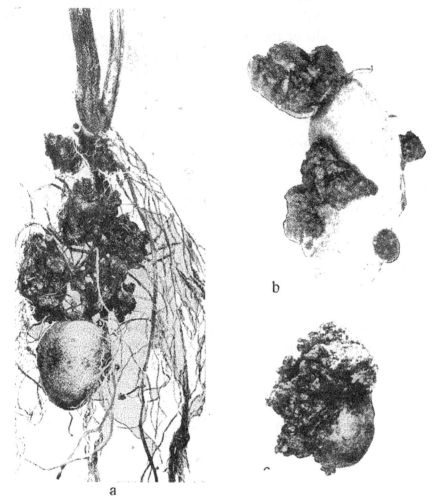

FIG. 18.—POTATO BLACK-SCAB—*Chrysophlyctis endobiotica.* (*a* AND *c*, FROM O. APPEL ; *b*, FROM T. JOHNSON.)

a, The underground parts of a sick plant ; *b* and *c*, diseased potato-tubers.

The resting spores might settle in the surrounding earth, thus infecting the same ; or if slightly infected potatoes are used for planting, these spores will mix with them.   Soil that is infected can retain the infecting power for six to eight years. The attack of the disease is more or less severe, all depending upon different sorts of potatoes. In England the following kinds have proved to possess a great power of resistance: " Snowdrop," " Main Crop," " Lang-worthy," " What's Wanted," and " Conquest."

The disease was noticed for the first time in Hungary in 1896. Since 1900 it has become prevalent in England, where it has spread to an alarming extent. Lately it has manifested itself in Scotland and Ireland, and in two districts in Germany ; during 1908 in the Rhine provinces and Westphalia, and in 1909 in Silesia and Newfoundland.

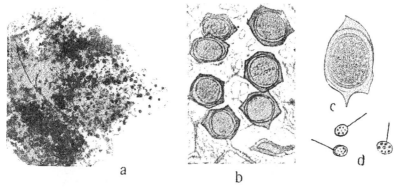

a
b
c
d

FIG. 19.—POTATO BLACK-SCAB. (*a, c,* AND *d,* FROM T. JOHNSON ; *b,* FROM K. SCHILBERSZKY.)

a, Section of a part of a tuber, with resting sporangia (slightly magnified) ; *b,* tissue with resting sporangia (highly magnified) ; *c,* a mature resting sporangium ; *d,* swarming spores.

Recently the opinion has been pronounced that the disease, which has turned up in the United Kingdom, is not quite identical with the one observed in Hungary in 1896, and the fungus has been given a different name—viz., *Synchytrium Solani* (*S. endobiotica*).

PROTECTIVE MEASURES.—(1) Totally destroy at once all diseased plants, and do not use potatoes taken from infected fields for seed, even if they have a sound appearance. (2) For a period of four to five years no potatoes should be planted in a field that yielded a sickly crop, although even only a small percentage. (3) Isolate a field that yielded a diseased crop, lest earth should be brought from it by means of implements, people, and animals, and mixed with

good soil. (4) If you are bound to use potatoes for planting that
you suspect are diseased, or if you must use a field that is infected,
you should prepare the soil with gas-lime, 4 to 5 tons per acre ;
and the potatoes that are to be planted should be sprinkled with
sulphur dust, 4 to 5 pounds per ton.   (5) It is best not to cultivate
potatoes at all in an infected field, but to use it for some other

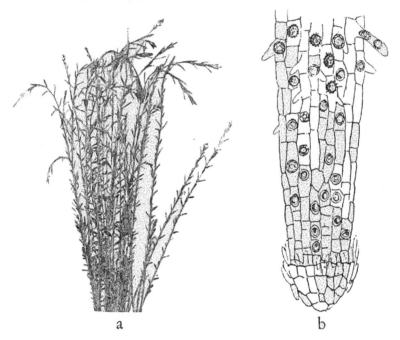

a   b

FIG. 20.—FLAX ROOT-BLIGHT—*Asterocystis radicis.*   (*a,* FROM E. MARCHAL ;
b, FROM V. DUCOMET.)

a, Plants in an early stage of the disease, the tops of the stems relaxed
    and drooping ; *b,* the top of a branch of the root, with fungi in the
    cells.

vegetable, after you have previously arrested the infection in the
following way : dress the soil in the month of March or April
with 2 pounds of gas-lime to each square yard, and plough this
down to a depth of 3 inches ; or mix in the month of May sulphur
powder with the earth (4 ounces per square yard).   (6) If potatoes,
suspected as diseased, are to be used for fodder, great care should
be exercised that no fragments of them find their way into the manure

or remain in the mangers, thus making sources of infection. It is best to boil such potatoes before using them as fodder.

Flax Root-Blight. (*Asterocystis radicis.*)

This disease appears in the early summer in sundry spots in the flax-fields. The lower leaves of the affected plants become yellow. The stem gets slack, with a relaxing and drooping top ; the ultimate root-strings become glassy and brittle. If the weather be rainy or the soil damp, the disease will quickly disseminate over the entire field.

The fungus that causes this disease (*Astero cystis radicis*) is located in the outer cells of the fine root-strings. It penetrates there while the roots are very young, and forces an entrance above the point of the root. This is easiest done from the thirteenth to the eighteenth day after sowing the flax. After the plants have reached their twenty-fifth day they are free from danger. The fungus goes from cell to cell. In the diseased cell the plasma body, which seems to embrace the plasma of both the cell and the fungus, looks like a sediment, sometimes filling the entire cavity of the cell. By degrees the plasma body is transformed into one or more masses of swarming spores, which gain an exit through an opening in the wall of the cell. The fungus subsists from year to year with resting spores, one or more in each cell. It can also live in the roots of other plants, such as cabbage, turnip, radish,

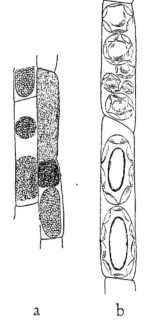

a   b

FIG. 21.—FLAX ROOT-BLIGHT. (FROM E. MARCHAL.)

a, Cells from the exterior parenchyma of the root, with fungoid bodies in the cells, in an early stage of development ; *b*, cells with resting spores.

lucerne, lettuce, onion, and others. This disease is known in Holland by the name of "Brûlure," and is also prevalent in Belgium and France, and has been noticed in Germany and Ireland.

No other protecting remedy is known for this disease than to avoid for a period of seven to ten years the cultivation of flax in a diseased field.   It has also been advised to pull up and burn every plant that shows an indication of the disease.

Amongst others belonging to the group of the Chytridiaceæ fungi may be mentioned :

Urophlyctis leproidea, or " Beetroot Tumour," which forms large tumours on the root of the beet either at its top or its middle part.

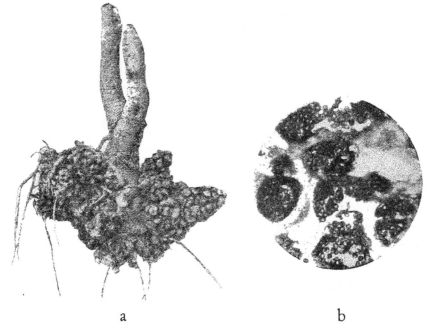

a   b

FIG. 22.—CROWN-GALL OF LUCERNE—*Urophlyctis Alfalfæ.* (FROM G. KORFF.)
a, Root of lucerne with tumours ; *b*, section of the tumour, showing cavities filled with resting spores.

These tumours have a short stem, and are irregularly intersected. When you examine them, they show dark pricks on the surfaces. The disease has been noticed in Algeria, France, and once in Sweden (1900) and in England (1905).

Urophlyctis Alfalfæ, or " Crown-Gall of Lucerne," causes in lucerne coral-shaped tumours on the neck of the root about the size of a pea.   The disease was first seen in America, then in Switzerland, Italy, Alsace, England, and Bavaria.

Olpidium Brassicæ, or "Seedling Cabbage Disease," attacks tiny cabbage-plants in the spring, especially if they grow crowded and are kept very moist. The disease manifests itself in such manner that the head of the root gets black and breaks, when, in consequence, the plants wither away.

O. Nicotianæ.—Possibly a variety of the previous one. It attacks young tobacco-plants.

O. Trifolii, which causes blister-like swellings on white clover.

Pyroctonum sphæricum, which kills wheat-plants, thus spreading yellow, dead patches over the fields; it is noticed in France.

CHAPTER IV

DOWNY MILDEWS—PERONOSPORACEÆ

THESE fungi are nearly all parasites. They possess a well-developed and richly ramified system of spawn (*mycelium*). The vegetative

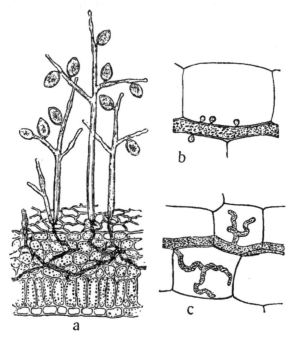

FIG. 23.—MYCELIUM, VEGETATIVE AND FRUGIFEROUS.

a, Intersection of a diseased leaf, with vegetative mycelium inside the leaf and frugiferous outside; *b,* vegetative mycelium thread, with short, globular haustoria (after *Cystopus*); *c,* a similar one, with long, ramified haustoria (after *Peronospora*).

(nutritive) part of the mycelium extends as a mesh within the host plant between the cells, and penetrates by means of small side branches (*haustoria*) into the cells. The fructificative (spore-producing) part of the mycelium appears outside the host plant as a grey mould. Examined through a microscope, this mould is seen to consist of numerous erect aerial threads or filaments, often repeatedly bifurcated, which bear egg-shaped breeding-cells, by which the fungus propagates during the period of growing. As a rule, the germination takes place in such a way that the contents subdivide into a number of swarm spores, which are liberated one by one, and extend a germinating tube or filament. Occasionally the breeding-cell sends out undivided one tube or filament, in that case acting as a *conidium*. For the continuance of the fungus until the coming year there are provided thick-walled resting spores, which later in the season develop within the affected cell-tissue, but frequently not before the same has begun to decay; these do not germinate until the winter is gone.

The attack of the fungus has usually this effect, that the cell-tissue very soon dies. In some cases, however, only a distortion takes place, while the host plant loses the power of producing ordinary blossoms or to yield ripe fruit.

Seedling Blight. (*Pythium Baryanum.*)

This fungus thrives in wet earth, especially in hotbeds, and proceeds from the soil to young plant germs, on which it especially attacks that part of the stalk which is under the seed-leaves (cotyledons). The plants become soft, bend down, and die. For this reason the disease has also been named "damping-off of seedlings." In moist air there grow out short fungus threads through the epidermis, and these cast off spherical spore-cases, the contents of which are diffused as a blister, and resolve eventually into numerous swarm spores. These spores are capable of immediate germination. The resting spores, intended to last through the winter, develop later in the decaying tissue, and remain in the ground. The fungus attacks plant germs belonging to many different varieties of plants, such as beet, barley, peas, lupin, clover, cabbage, spurry, sinapis, asparagus, stock, and others. Recently it is spoken of—especially in Germany—as a parasite on beet-plants. Sometimes it will attack older portions of plants, as potato-tubers and stalks, heads of asparagus, etc.

PROTECTIVE MEASURES.—(1) Mix into the upper layer of the earth in hotbeds, where plants are germinated, fine sand or coal dust, and keep them sufficiently moist and warm during the early stage, while

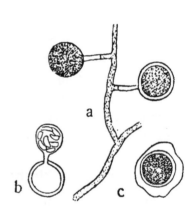

FIG. 24.—GERM BLIGHT IN BEET— *Pythium Baryanum.* (FROM W. BUSSE.)

a, Diseased plants; *b,* sound plant (at the same age).

FIG. 25.—PYTHIUM BARYANUM. (FROM E. ROSTRUP.)

a, A filamentous tube, with two spore - cases ; *b,* a spore - case ready to develop swarm spores ; *c,* a resting spore.

the stalk beneath the cotyledons still keeps on growing. (2) If the disease appeared in a hotbed during the previous year, then new soil should be brought and all frames and glasses thoroughly cleansed. (3) If there be reason to suspect that the beet-seed contains any infection, then the seed-bunches should be kept moist for two to three days, and then steeped for two hours in a solution of sulphate of copper to a strength of 1 per cent. to 2 per cent. before they are sown.

Potato Blight. *(Phytophthora infestans.)*

This is the best known of all the different diseases to which the potato-plant is subject, and is generally referred to as " Potato Disease." It shows its presence by the appearance of large dark spots on the leaves during the time the plant is blooming or shortly

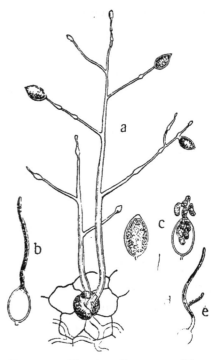

FIG. 26.—POTATO BLIGHT—*Phytophthora infestans.* (FROM E. ROSTRUP.)

FIG. 27.—POTATO BLIGHT. (FROM E. ROSTRUP.)

a, Two branches of a filamentous tube protruding from a cleaving, and carrying breeding-cells; *b,* a breeding-cell germinating with sprout; *c,* two similar ones, just ready to discharge swarm spores; *d,* a swarm spore; *e,* a germinating swarm spore.

after. The spots increase daily both in quantity and size. Soon all the leaves wither away and the stalks are almost bare. Meanwhile a fetid and rotten smell emanates from the potato-field.

When examining a diseased spot on a leaf you will find the lower side covered with a dirty grey stuff or blight. At first this blight is evenly distributed over the whole spot, but as the spot gets older its centre, which was first diseased, dies, and as there now remains no nutrition for the fungus, the blight disappears from the centre, and is conspicuous only at the edges.

This dirt-grey blight is composed of hanging, ramified, colourless fungoid threads, which bear numerous egg-shaped breeding-cells. While germinating they act as a rule like spore-cases (*sporangia*)— *i.e.*, their contents arrange themselves into a varying number of four to sixteen separated plasma bodies, that pass out through an opening at one end of the spore-case. These bodies are provided with two fine vibratile cilia, which give them a gyratory or vibrating motion. After about half an hour this movement stops, the cilia drop off, and the spores begin to germinate. Sometimes it may, however, happen that the breeding-cells, which are cast off from the ramified mycelium threads, directly sprout a germ-tube, thus acting as an ordinary spore or conidium.

The extension of the disease over the potato-field depends to a considerable degree upon the state of the weather at the first outbreak of the disease and after. If the weather be damp, especially foggy, the disease will spread very rapidly — within one to two weeks the whole field is black. But if the weather be quite dry at this period, the propagation is tardy, and may even stop altogether, at any rate for some time.

It has occasionally happened that potatoes planted in hotbeds in January or February have shown the disease by the middle of April, just when the tubers were forming, and within a few days the plants have withered away.

When the leaves have thus died all over a potato-field, it will not be long before the new tubers begin to show ominous spots, especially those nearest to the surface of the earth. These are mostly of a brown colour, and penetrate more or less deeply into the potato. The way in which the disease reaches the tubers has generally been considered to be through the breeding-cells, which have dropped on the ground and have been washed down into the soil by the rain. It is true enough that such a surmise has been established by several experiments undertaken in 1880 and later on, both in Denmark and England. In these experiments spores were strained through layers of sand of various depths. It has also been tried with success to protect the tubers against serious attacks of the fungus by means of mounds arranged high round each plant.

Against this simple explanation of the dispersion of the disease we have several experiments of a later date. At these trials it proved next to futile to render a sound potato sickly by means of diseased leaves. And in those cases where disease developed, it proved to be not potato blight, but dry-rot, a disease which, as previously described, is caused by bacteria. If these latter results should be further confirmed, there remains no other way of explaining the access of the disease into the tubers than to accept the conjecture that it reached the tubers in the same manner as it came to the leaves.

But, then, how *does* the disease first get into the leaves? This question is as yet unsolved, and so is also another query : In what way does the fungus of potato blight hibernate? Several different hypotheses have been brought forward, but none of them have been definitely proved. The only thing we are sure of in regard to this fungus is, that it does not show a natural stage with resting spores as most of the other downy mildews do. At least, so far, all researches in the open for such a stage have proved futile.

The year 1845 is generally considered as the first year of the appearance of the disease in Europe, when an account was given of the fungus and its scientific name settled. The same year it was prevalent over a considerable portion of the Continent as a very destructive pest. The extraordinarily rapid spread of the disease has been considered as arising from the weather conditions during the summer and autumn of that year. But a quite satisfactory explanation of the phenomenon has scarcely been found. Equally remarkable and unexplained is the sudden appearance of this pest in North America.

Experience has proved that various kinds of potatoes are more or less severely attacked by the disease. This observation has been of service as a protection against the disease. But all the vicissitudes which a planter is subject to in this respect are fairly well known. Even if certain sorts of potatoes have once shown resistance, it by no means follows that they always will do so. On the other hand, it is not merely this point that has to be considered by the potato planters, but also other qualifications, such as rapid growth, prolific nature, taste, degree of starch, and so forth.

A sort that still keeps its good record is the old, well-known "Magnum bonum." As valuable new sorts might be mentioned "Königin Carola" (early food potato), "Up-to-date" (somewhat early food potato), and "Märcker" (rather late factory potato).

The same fungus has also been noticed in several other plants more or less related to the potato, such as *Solanum etuberosum, S. caripense, S. stoloniferum, S. utile, S. Maglia, S. verrucosum, S. Dulcamara, S. laciniatum, S. marginatum, S. tuberosum × utile, S. Lycopersicum, Anthocercis viscosa, Petunia hybrida, Schizanthus Grahami, Datura Metel,* and others.

For some time it was thought that the same fungus was the origin of the two diseases which late in the autumn affect the potato, and are commonly known as "wet-rot" and "dry-rot." But we have previously learnt that these diseases are caused by bacteria.

PROTECTIVE MEASURES.—(1)· Use only sound seed. Potatoes gathered from a diseased field crop should not be used for planting, even if they look comparatively sound. We have every reason to believe that it is by means of slightly diseased tubers and by the living germs concealed in them that the disease is retained from one year to another. (2) Select a field, the soil of which is appropriate for culture. The best would be in a dry locality with well drained sandy soil or clay strongly mixed with sand. (3) Select such kinds of potatoes as will have the greatest resistance, and let this selection be guided by previous experience gained in the locality. (4) Avoid cultivating potatoes year after year in the same soil. (5) Do not fertilize the field with manure from the farmyard, especially fresh manure, immediately before planting the potatoes. (6) Bank up the potatoes high. In that case the earth should be piled so high that the topmost tubers are 9 centimetres (about 3½ inches) down, if in sandy soil, and 10 to 13 centimetres (4 to 5 inches) if the soil consists of clay. This arrangement with mounds requires a wider distance between the rows than the customary one—viz., on sandy soil 80 centimetres (about 2½ feet), and on harder soil 95 centimetres (about 3 feet). (7) Sprinkle the field at the time the spots first begin to appear (or one week in advance in accordance with experience from previous years) with some fungoid-killing fluid, as, say, 1 per cent. Bordeaux mixture, and sprinkle it again after three to four weeks. During normal years it has proved best in Southern Sweden—and the case might most likely be the same in Southern England—to sprinkle early potatoes for the first time during the first half of July, and medium early and late potatoes during the latter half of the same month. The effect of such a sprinkling will be that the leaves remain green about one to five weeks longer than they otherwise would, and also that not only the number of sound tubers will increase, but the total weight of the whole crop will be higher. (8) Do not take the potatoes up from the ground too early, while the surrounding atmosphere is still filled with infectious germs from withering leaves and stalks, which as yet are not quite dead. (9) Store the potatoes in a dry condition, thus preventing the outbreak of disease during winter.

Beet Mould. (*Peronospora Schachtii.*)

Both fodder and sugar beets are subject to this disease. While young, the diseased leaves are pale, with curled and rolled-back edges. When older, they are pulpy and swollen, and covered on the under side with an ashy-coloured or yellowish mould.

This grey mould consists of repeatedly branched or forked fungoid threads or filaments, which at the points bear egg-shaped breeding-cells, the source of the disease. The attacked leaves die prematurely. In old withered leaves there occasionally develop resting spores, which are thick-walled and hibernate.   Sometimes rotten spots appear on the roots.   It is alleged that the fungus can exist from one year until the next as a sterile thread-mesh in the top of hibernating seed-beets.

This disease is known in Denmark, Germany, and France, and occasionally is very destructive, as was the case in Saxony in 1894.

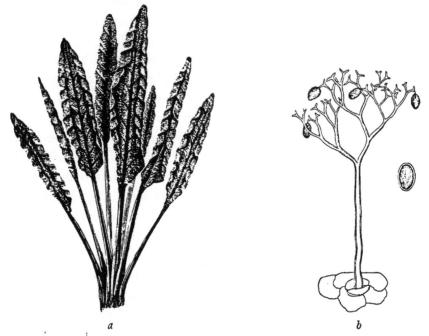

a   *b*

FIG. 28 —BEET MOULD—*Peronospora Schachtii.*   (FROM E. ROSTRUP.)

a, The top of a beet-plant (reduced size) with the edges rolled backwards ; *b,* re-
    peatedly forked or bifurcated fungoid threads or filaments, with breeding-
    cells.

PROTECTIVE MEASURES.—(1) Do not take seed-beets from a dis-eased field.   (2) Try early to sprinkle the infected beet-fields with a 2 to 3 per cent. Bordeaux mixture.

Among other downy moulds injurious to agricultural plants can be mentioned :

Peronospora Viciæ on peas, fitch or vetch, and other plants related to them.   It forms grey-violet blight patches close to each other, or also one continuous blight over the under side of the leaves. This blight consists of repeatedly branched or bifurcated threads, giving off breeding-cells at the points. Within the diseased leaf-tissue the hibernating resting spores are developing. It frequently happens that the harvest of fitch or vetch, etc., is greatly reduced through this attack.

P. trifoliorum on both cultivated and wild clover, and on several other related leguminous plants, such as lucerne, lotus, and

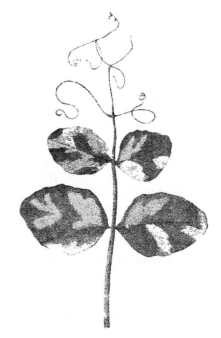

others. It produces larger or smaller pale patches on the leaves, which have often the shape of broad, cross-going tapes.

The patches are on the under side of the leaves, covered by a grey or pale - lilac coloured mould, with almost globular breeding-cells on the tops of the filaments. Its ravages are worst on lucerne, especially if this has come from America. It has been noticed as a practically different degree of disease from different seed, even although the plants have grown side by side. This fact would indicate that the disease is spread by means of the seed.

FIG. 29. — BLIGHT ON PEAS—*Perono-spora Viciæ*. (FROM E. ROSTRUP.)

Resting spores are developed quite sparsely and have been noticed in the tissue of the stipule.

P. parasitica on turnip, swedish turnip, rape, sinapis, and other both cultivated and wild cruciferous plants. The fungus forms white, flour-like coverings on stem, leaves, and the blooming sprout. Often the attacked parts become quite distorted.

P. obovata on spurry ; the leaves turn mouldy.

Plasmopara nivea on carrot and parsnip. The under side of the leaves is covered by a snowy-white mould.

PROTECTIVE MEASURES AGAINST SEVERAL OF THESE DISEASES.— (1) Do not plant in earth which has produced a sickly harvest for several years the same kinds of plants, neither any related to them that might be susceptible to the disease. (2) Do not use seed from

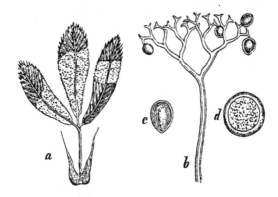

FIG. 30.—BLIGHT ON LUCERNE—*Peronospora trifoliorum.*
(FROM E. ROSTRUP.)

a, A leaf of lucerne with pale blight patches; *b*, a forked thread with breeding-cells; *c*, a breeding-cell; *d*, a resting spore.

diseased plants. (3) If possible, try and sprinkle the field in the early summer with 2 to 3 per cent. Bordeaux mixture.

Cabbage White Rust. *(Cystopus candidus.)*

This fungus, known as "White Rust," differs from the previously described downy moulds in that the fertile filaments of the mycelium do not become conspicuous outside, but are hidden under the cuticle, the latter being of a shining white appearance. In this case they are short, undivided, crowded threads, which at their tops bear breeding-cells arranged moniliformly, or like bead-strings. Finally, however, the epidermis breaks, and the white spore meal becomes free. Inside the diseased tissue there develop hibernating spores.

Such a white rust (*C. candidus*) is prevalent on a number of wild cruciferous plants, as *Capsella, Thlaspi, Sinapis,* and others, and also

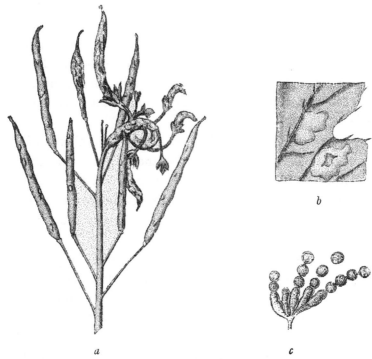

a   *c*

FIG. 31.—WHITE·RUST OF RAPE—*Cystopus candidus.* (FROM O. KIRCHNER AND H. BOLTSHAUSER.)

a, Immature fruit with fungoid swellings ; *b,* two fungoid swellings, one of them broken ; *c,* fungoid tubes with breeding-cells.

occasionally on cultivated plants related to them, as cabbage, swedish turnip, turnip, rape, and others. However, this disease rarely causes any considerable loss to agricultural plants.

CHAPTER V

SMUTS—USTILAGINACEÆ

THESE fungi—often named *Bunt*—are genuine parasites, which expand their tiny mycelium for some time through the whole plant without hindering its normal development to any visible extent. They reach up to the blooming stalk, where the formation of the spores takes place, and a thorough devastation sets in. The spores appear as a brownish-black mass of dust in long stripes on the stalks and blades of the leaves, or they wholly occupy the generative organs of the plant.

The spores are spherical, and provided with a thick wall, which often has wart or mesh-formed protuberances. As a rule the spores appear separated from each other; occasionally a few may be seen together. They either germinate at once, or they act as hibernating spores, in which case they are said to retain their power of germination as long as ten years.

When the spores germinate, there appears either a germ thread, that at once penetrates into the host plant (if the germination has taken place on the surface of the plant), where immediately a spawn is formed, or otherwise there develops a short and thick tube formation (*basidium*), which bears spores (basidic spores). If the germination takes place in a specially prepared nourishing fluid, as fertilizing extract, or the like, then a rich formation of bud-cells, similar to yeast fungi, will develop. As long as nourishment is available, this formation can go on indefinitely. From such cultivation infection can be brought on very tiny parts of a plant which is susceptible to the disease.

44

Stinking Smut. (*Tilletia.*)

The spores are simple, warty, and seldom smooth on the surface. At germination they develop a short basidium, which at its summit has a ring of thread-like basidic spores, often united at their middle part in pairs, and either give off germinating threads, or produce bud-cells, which germinate.

Stinking Smut of Wheat. (*Tilletia caries* and *T. levis.*)

This smut, also called "Stinking Bunt," does not indicate its existence when first attacking wheat-plants. These appear to be healthy until the time comes when the seed begins to ripen. At this time the diseased plants become ominously conspicuous through their dilated ears and their shorter and thicker corns. As a rule every ear in a plant and every corn in an ear are diseased. Only occasionally sound and sickly corns are found in the same ear. At maturity the affected corns are grey outside, while inside they are filled with a brownish-black mass of dust. They have a fetid smell, peculiar, penetrating, and disagreeable; hence the disease is called "Stinking Smut."

This disease is brought about by either one of the two fungi: *Tilletia Tritici* and *T. levis.* The former has warty spores, the other has spores with a smooth surface. It is principally *T. Tritici* that is the offender.

The husk of the smut-corns does not break of its own accord, but these corns are brought together with the sound ones into the granary. At the threshing they are more or less crushed, and the black spore dust gets free. Spores then fasten easily on the healthy corns, especially on the hair pencil at the summit. Infected wheat gets a blackish-grey colour and smells badly, as well as the flour prepared from it.

By means of infected corn the disease is propagated to next year's crop. When the wheat-corn grows, then the smut-spores grow on its husk. These form a short tube, which has at one end a ring of long, narrow spores, that either directly, or through separate bud-cells, send germinating tubes into the young wheat sprigs. Such infection is called "germ infection." If the fungus once gains an entrance into the tender plant, then it will spread through the

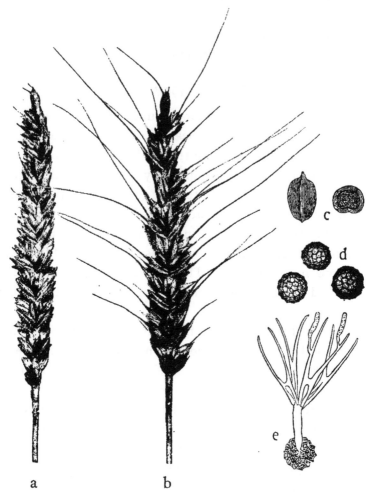

a   b

FIG. 32.—STINKING SMUT OF WHEAT—*Tilletia Caries*. (*a* AND *b*, THE AUTHOR.)
a, Smut-ear of unbearded wheat; *b*, smut-ear of bearded wheat; *c*, two smut-corns, one whole and the other cross-cut; *d*, three spores; *e*, spore that has germinated.

entire plant down to the ovule, within which it develops its dust-like mass of spores. The spores retain their germinating power for years.

Different sorts of wheat, especially spring wheat, are susceptible in another way. But even in the same sort the effect of the attack can vary during different years. Of special importance in this respect are the climatic conditions at the sowing and subsequently. A low temperature at this time will delay the germination of the wheat, thereby prolonging the period under which it is exposed to infection. Hence a late sowing of autumn wheat and an early sowing of spring wheat promotes the ravages of the disease.

It has sometimes been feared that smutty wheat, given as fodder to the stock, may also convey disorder. But to judge from experiments with horses, cattle, pigs, hens, and pigeons, this is not the case. Neither has any lowering in the general condition of health taken place, nor have any sickly formations been found in their inner organs.

PROTECTIVE MEASURES.—1. *Take the grain for sowing from a field where no disease has appeared during the year.* Should this prove to be impossible, the following precautions might be observed:

2. *Wash the Sowing-Grain.*—If the sowing-seed is mixed with smut-corns, they must be separated by pouring the grain slowly into a vessel containing water, and thoroughly stirred. The smut-corns float, and are skimmed off.

3. *Swamp the Sowing-Grain.*—Since, even in spite of the washing, there are always to be found smut-spores on the corns, especially at the top, it will be safest to swamp the grain. Several methods may be chosen:

(a) *Steeping in a Solution of Copper Vitriol.*—Half a kilogramme of copper vitriol is dissolved in 100 litres of water, or say about 1 pound to 20 gallons. The vitriol is put in a cloth bag, which is kept in the water until all is dissolved. The solution is then stirred up carefully, and the wheat poured in in such quantity that the fluid is about 4 or 5 inches above the wheat. The seed is stirred repeatedly. After twelve to fourteen hours the fluid is poured off and the grain poured out on the floor of the granary, that previously has been washed with a similar solution. Then the seed is sprinkled with a lime solution prepared from 1 kilogramme of slack lime and 100 litres of water, or say about 2 pounds to 20 gallons. Finally, the seed is spread to dry. As soon as sufficiently dried, it is sown. The sacks in which the grain is transported should previously have been soaked for at least twelve hours in 2 per cent. vitriol solution.

(*b*) *Crystallising or Candying with Bordeaux Mixture.*—In a vessel 2 kilogrammes of copper vitriol (copper sulphate) is dissolved in 50 litres of water, or say about 4 pounds to 10 gallons. In another vessel lime-wash is prepared in such a way that 2 kilogrammes of new burnt lime (stone lime) is so moistened with water that it becomes a white powder. To this 50 litres of water is added to the solution and well stirred. Both the solutions are then mixed together in equal strength. The Bordeaux mixture prepared in this way should have a pretty blue colour, turn red litmus-paper blue, and deposit a blue precipitate at the bottom. In this solution a wicker-basket, lined inside with coarse linen, is submerged. Then the grain is poured into the basket, stirred up a few times, while the smut-corns and spores that are floating up to the surface are being skimmed off. After ten to fifteen minutes the basket is taken up and the grain spread out to dry. By such procedure the corns get candied with a thin crust; this will remain on them when they are sown, and will at the germination act as a spore-killer (fungicide).[1]

(*c*) *Steeping in a Formalin Solution.*—In a 36 to 40 per cent. water solution of formalin $\frac{1}{4}$ to $\frac{1}{2}$ litre is mixed with 100 litres of water (say, $\frac{1}{2}$ pint to 1 pint to 20 gallons). Into this solution the seed is poured, all that floats up to the surface being skimmed off, while it is stirred repeatedly. It is then left to stand for half to one hour; then the fluid is poured off, and the seed spread to dry. For such a steeping on a large scale there is constructed a special apparatus—" Dehne's Disinfection Apparatus." This is, however, applicable only for seed that is free from whole smut-corns. With this apparatus one can sodden 1,000 kilogrammes, or about 1 ton of seed, in half an hour.

---

[1] The disadvantage in using copper vitriol is that the power of germination will be somewhat diminished, especially if the seed be machine-threshed. Another trouble is that if there be any seed left over after sowing, such cannot be used for foodstuff, as it is poisonous.

(d) *Ceres Treatment.*—The Ceres powder contains as fungicide principally sulphurated potash, and also substances calculated to promote a rapid germination of the corn and a quick development of the germ-plant. The powder is dissolved in cold water, 1 kilogramme powder to 350 kilogrammes seed, say about 2 pounds to 7 cwt. The grain-heap is sprinkled repeatedly with the solution, stirred up thoroughly, and allowed to remain thus soaked from ten to twelve hours ; then it is spread out thinly to dry. The Ceres treatment does not exterminate the disease altogether ; it only has a diminishing effect. If it is adhered to for several years, the percentage of the disease will be reduced year by year, until finally there is not much of it left. Seed treated with Ceres should not be sown sooner than two to three days after the treatment.

(e) *Warm-Water Treatment.*—Under this treatment the seed is submerged in water of about 56° C., after having first been placed in a lined wicker basket or in coarse sacks. Should the temperature sink at the submersion, then warm water should carefully be poured in until the temperature of the water in the vessel is the required one. As an alternative, before the submersion, dip the basket or the bags containing the seed repeatedly in water of 40° to 45° C. The seed is kept submerged in the 56° water for about ten to fifteen minutes, and is then spread out to dry quickly in the air.

(f) *Treatment with Warm Air.*—Then the seed is exposed for a quarter to half an hour to air of 60° to 65° C., and this sort of treatment has latterly been tried with success. For this treatment a special apparatus is required for drying. Such an apparatus, named " Selecta," has been introduced into the market by the firm, Select Grain Machine Company, Berlin, W. 35.[1]

Amongst the different methods of soddening, the warm-water treatment is now considered the best and most reliable one. As, however, its proper procedure requires more time and labour than planters in general feel inclined to bestow upon it, various means have been adopted in different countries to render

[1] Seed treated with formalin, warm water, or warm air, can be used for other purposes, if all is not needed for sowing.

the method more practicable.   In Denmark several dairies have been induced to do the work for payment, as these establishments have easy access to warm water, necessary vessels, and so forth, and also have a working staff accustomed to work with the use of the thermometer.   In Germany there have been constructed various apparatus, as " Arnim-Schlagentin's," by means of which the seed is passed through warm water ; and " Appel and Gassner's," in which the warm water is passed through the seed.   With the latter, 50 hectolitres (equal to 135 to 140 bushels) of seed can be sodden in one day.

4. Take care lest the threshing-machine bring infection, as it might previously have been used for infected seed.

Among other forms of " stinking smut," that destroy grasses by filling them with a greyish-brown dust-mass, may here be mentioned : **Tilletia Secalis,** on rye, noticed in Germany, Austria, and Italy ; **T. Lolii,** on rye-grass (*Lolium*), in Germany and Denmark ; **T. decipiens,** on *Agrostis*, in Germany, Denmark, and Sweden ; **T. Holci,** on soft-grass (*Holcus*), in Belgium and Denmark.

Other forms belonging to this sort develop their masses of spores in long wound-stripes on the straws and leaves of grasses.   Such a one is **T. striæformis,** which attacks a great number of different grasses, as *Poa, Holcus, Festuca*, and others.   This disease is known in most of the European countries.

Loose Smut.   (*Ustilago.*)

The spores are simple, either warty or smooth.   At germination there is formed a short basidium, which divides into joints, each one developing a basidium spore.   The spores are seldom directly germinating.

Loose Smut of Wheat.   (*Ustilago Tritici.*)

In contrast to the stinking smut, the loose smut becomes conspicuous as soon as the wheat begins to form ears, and the corn husk that covers the dust-mass cracks immediately, when the spores are scattered by the wind. Soon there will remain of the whole ear nothing but the bare rachis, partly black by means of smut-spores that remain at the joints.

The disease is caused by a special fungus, *Ustilago Tritici*. For a long time it was thought that the same fungus brought on the loose smut disease on wheat, barley, and oats ; recent researches, however, have proved that this is not the case, but that the loose smut of different seed belongs especially to each one of them, and does not infect the others.

The spores of the loose smut are globular and unicellular, with small warts on the surface. They can germinate at once, but do not retain their power of germination longer than a few weeks ; consequently they cannot infect sound corn in the same way as the spores of the stinking smut.

The spores of the loose smut diffuse simultaneously with the blooming of the wheat. The spores stick to the viscid stigmas of the pistil, and germinate there quite readily, just like pollen, and send their germ-tubes right through the style down to the developing seed, thus bringing on the infection. This form of infection is called "blooming infection." A seed-rudiment infected in this way does not show the infection during the first stages of development. It grows and ripens just like any ordinary wheat-corn, and cannot in the cleansed stock of grain be picked out from the sound ones, although the infected corns do not seem to equal the average size of the sound corns ; but in the next crop the latent disease becomes conspicuous.

If the infected seed is sown, then the plants become sick with the smut. The ears on the formation of the corn are filled with a black mass, which is at once scattered by the wind, and continually provokes fresh infection. The conditions of the weather during the period of blooming will influence the dispersion of the loose smut. If heavy rain showers occur when the smut-ears are about to let out the black mass of spores, then a number of the spores—possibly most of them—reach the ground and germinate there, thus never reaching the styles of the blossoms ; but if the weather be dry and calm during the blooming period, then the blossoms are subjected to infection.

As long as the corn is dry, the fungus that has infected it is in a

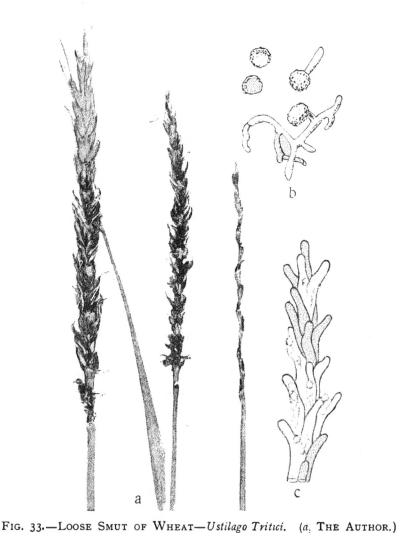

FIG. 33.—LOOSE SMUT OF WHEAT—*Ustilago Tritici.* (*a*, THE AUTHOR.)

a, Three smut-ears ; one of them has lost both smut-corns and awns, and the wind has left only the bare ear-stalk ; *b*, spores, before and after germination ; *c*, style of a blossom infected with smut.

stage of perfect rest, and can in no way be influenced from without ; but if the corn be kept for several hours in lukewarm water, then the fungus changes into a stage of germination, and becoming influenced by higher temperatures, is in consequence open for treatment with warm water or warm air.

When the wheat ultimately ripens, there is hardly any dust of this sort of smut ; but even if it were there, it could not penetrate into the ripe, sound corn. Consequently a plant grown from such a corn is sound, as this sort of smut cannot produce germ infection.

PROTECTIVE MEASURES.—(1) Do not take sowing-seed from a field where during the blooming period loose smut has been prevalent, as presumably many blossoms have been infected, and seed of such blossoms produces smut-plants in the following crop. (2) If the sowing-seed comes from an infected field, or belongs to a kind of seed that is known to be easily susceptible to the disease, then sodden the sowing-seed with warm water or use warm air in the following manner : Steep the sowing-seed first in lukewarm water of a temperature of 20° to 30° C. for four to six hours, and submerge it in warm water of 50° to 54° C. for twenty minutes, or expose it to warm air of 55° to 60° C. for thirty minutes. The percentage of the disease has thus in numerous experiments sunk to next to nothing. (3) Try to prevent the infection during the blooming period. On smaller experimental plots this can be done by closely inspecting the wheat day by day during the formation of the ears, and pulling up by the root and burning every plant that is setting smut-ear. This is continued until it is certain that every plant has produced at least one ear. In the same way the private planter can also procure a sound wheat stock if he reserve in his plantation a special little experimental field far away from the rest of the wheat-fields. In this the weeding out of the diseased plants can be done, and, by taking special care during the period of the first formation of the ears, he may finally reach his goal, and be in possession of a perfectly sound sowing-seed for all his fields.

Loose Smut of Barley. (*Ustilago nuda.*)

This smut corresponds with the loose smut of wheat ; it appears when the barley is forming the ears and beginning to bloom. The

FIG. 34.—Loose Smut of Barley—*Ustilago nuda.* (The Author.)
Three smut-ears, the third having only the rachis left, the wind having carried off awns and smut-corns.

dust-mass is at once scattered by the wind. Several spores are thus brought into the blossoms of sound ears. Some of them stick to the young stigmas of the pistils. There they germinate, sending germ-tubes down to the developing seed, which thus gets infected. The seed of the infected blossom develops like an ordinary corn-seed, but the infection is latent, and next year a diseased plant will develop from this previously infected corn. This disease is caused by a fungus named *Ustilago nuda*. The spores are globular, with a warty wall. At their germination they develop a long, jointed, and ramified germinating filament that infects the pistil.

Some sorts of barley are more susceptible to this disease than others, and this is evidently owing to the difference that the pistil of one sort is less protected from infection than is the case with another sort. In the southern part of Sweden the populace evince great preference for the "Hannchen" barley, but this is very susceptible to the disease. If the weather be dry during the blooming period, then the spores can float about, while strong showers of rain wash them down to the ground, where they can cause no harm.

PROTECTIVE MEASURES.—The same as for loose smut of wheat.

Covered Smut of Barley. (*Ustilago Hordei*.)

It is only lately that we have learned to distinguish between two different sorts of smut of barley. The covered smut differs in many important ways from the loose smut. Barley ears affected by covered smut are during the blooming period not very much unlike the sound ones. But a few weeks later the smut-ears catch the eye through their darker hue; their tiny ears are distorted into a broad, swollen, tripartite formation, the middle tooth corresponding to the middle corn and the side teeth with the side blossoms.

The husk that covers the smut remains as a silver-grey pellicle, quite perfect until the barley ripens, only showing a few unimportant cracks. Thus no infection of the blossoms takes place, as the spores do not spread while the barley stands on its root. The covered smut-ears are brought with the sound into the granary, and break only at the threshing. The spore-stuff is then spread to the sound corns, and, should the same still remain fit for germination when this barley is sown, then the germinating sprout becomes infected, and the growing plant produces smut-ears. In this case the germination infection takes place in this way.

The disease is caused by a special smut, named *Ustilago Hordei*, with larger and more edged spores than those of the previously described smut; their cell walls are smooth. When the spores germinate, they develop a short germ filament that bears bud-cells.

PROTECTIVE MEASURES.— The same as against stinking smut of wheat. Of 36 per cent. formalin, there is taken 1 kilogramme to 225 litres of water—say about 2 pounds to 45 gallons—and the steeping should last for ten to fifteen minutes, after which the seed is left untouched in a heap for seven to eight hours. Sulphate of copper reduces the power of germination.

Loose Smut of Oats.
(*Ustilago Avenæ.*)

This, the more common of the two sorts of smut to which oats are subject, appears like the loose smut of wheat and barley. As soon as the oats begin to form

FIG. 35. — COVERED SMUT OF BARLEY— *Ustilago Hordei*. (THE AUTHOR.)

ears the smut becomes conspicuous. The attacked ears are narrower than the sound ones, and the bristles are less dilated. Sometimes all the small ears in a panicle are affected, but it might also happen that sound ones are mixed with them, especially at the top.

Those of the spikelets that are entirely destroyed are of a globular shape, forming one homogeneous mass of dust. Other spikelets that are only partly attacked retain their natural shape better, and only their lower part is filled with smut.

When the smut-ears first appear, the husk of the corn is more or less cracked, and the smut is ready to diffuse. This diffusion lasts in their case longer than with wheat and barley—viz., from the commencement of the blooming until the period of ripening. By harvest - time the wind has carried off most of the spores.

In spite of the early dispersion of spores, it appears in this case that a blooming infection seldom, if ever, takes place. The disease is infecting the plant during the blooming period in such a way that spores settle in the open oat blossom, outside the seed-rudiment, and are retained there during the growth of the seed between this and the tightly closing husks. Later on the disease also spreads through spores that have stuck outside the growing oat-corn.

FIG. 36.—LOOSE SMUT OF OATS—*Ustilago Avenæ*. (THE AUTHOR.)

Three smut panicles, from one of which the wind has carried off the awns as well as the smut-corns; only the stalk of the panicle is left.

This disease is caused by *Ustilago Avenæ*, which has globular spores with a warty surface. When these germinate they usually develop a short germ sprout, from the top of which, as well as from the joints, bud-cells fall off, and in their turn germinate and infect the tender plant. The germination of the spores thrives best when the weather is warm and the sowing takes place in the spring. The spores retain their power of germination for a long time, certainly for several years. All infection with this fungus is germ infection.

It is not unusual for 30 per cent. of the crop to be lost through this smut—in fact, cases are known where the destruction has been as high as 60 per cent. And very seldom is an oat-field quite free from it. Thus it is wise to beware even of a slight attack, as this, under conditions that might be favourable to the development of the ailment, may lead to a great destruction the following year. Different sorts of oat are susceptible in different degrees. In Canada (Guelph, Ontario) there is an early sort of oat—" Early Ripe "—that is almost immune. The loose smut of oats depends greatly upon climatic conditions. Dry and calm weather during the blooming period is auspicious for the spores while getting into the blooms, and damp, warm weather at the time of sowing promotes the infection of the sprouting corn. Both of these conditions in conjunction may bring about a serious outbreak of the disease. This explains the variations of crops in different years.

PROTECTIVE MEASURES.—The same as for stinking smut on wheat and covered smut on barley.

The washing of the sowing-seed is in this case futile, as the moisture would not reach those spores that are closed up between the husk and the kernel of the oat-corn, where they penetrated during the blooming period. Of the different methods of steeping, that with sulphate of copper should not be adhered to, as the germinating power of the oat would thereby be considerably checked. If formalin solution is used, then 36 per cent. formalin should be taken in the proportion of 1 kilogramme to 300 litres of water (say about 2 pounds to 60 gallons), and the steeping last for ten minutes, after which the seed is left untouched in the heap for seven to eight hours.

Covered Smut of Oats. (*Ustilago Kolleri.*)

The oat panicles, which are affected by this smut, show at first nothing different from the sound ones, neither with regard to the

FIG. 37.—COVERED SMUT OF OATS— *Ustilago Kolleri*. (FROM O. APPEL.)

FIG. 38.—USTILAGO PERENNANS. (THE AUTHOR.)

panicles themselves nor to the spikelets. Later on, towards the time for ripening, the black mass of spores becomes conspicuous when showing through the husk of the corn. The husk keeps whole, at any rate until the ripening time; then when the grain is threshed the spores are let loose and stick to sound oat-corns. If they are let alone upon these, then the infection will be conveyed to the tender germ-plant.

This disease comes from a smut-fungus, called *Ustilago Kolleri*, or *U. levis*, different from the previous one in that the spores are on an average somewhat larger, more edged, and smoother on the surface. They germinate in the same manner as the loose smut on oats.

PROTECTIVE MEASURES.—The same as for the loose smut of oats.

Amongst other forms belonging to this sort of smut-fungi, which attack and destroy the fruits of grasses, may be noticed: **Ustilago perennans** on *Avena elatior*, which is thought to hibernate in the rhizome of the host plant, and renders the fodder poisonous to cattle; **U. bromivora** on brome-grass (*Bromus arvensis, B. mollis*, and others); both of them, especially the second one, are not uncommon.

Beside these there exist several forms of loose smut, which develop their spores on leaves and stems of different grasses. One of these, **Ustilago longissima,** forms on the leaves of *Glyceria aquatica* long, parallel, open stripes, filled with spores of an olive-coloured hue. The attacked sprouts as a rule develop no ears. Cattle eating fresh sprouts that are smutty, become ill, and may even die. Several such cases occurred in 1899 in the southern part of Sweden. On one estate twenty-four cows became ill. The symptoms of the sickness set in about one hour after the fresh grass had been consumed, and appeared as diarrhœa, stiffness, and a lowering of the temperature of the body. Most of them gradually recovered, but after that they never touched the grass. On another estate in the same locality three cows became ill in the same way, and so severe was the attack that the animals had to be killed. It is only the fresh grass that shows these poisonous qualities. When dried it is evidently harmless. Another form, **U. grandis**, brings on swellings of the thickness of a finger on the topmost joint of the stem of reed, *Phragmites communis ;* these swellings are filled with a black mass of dust. The affected stems do not develop any ears.

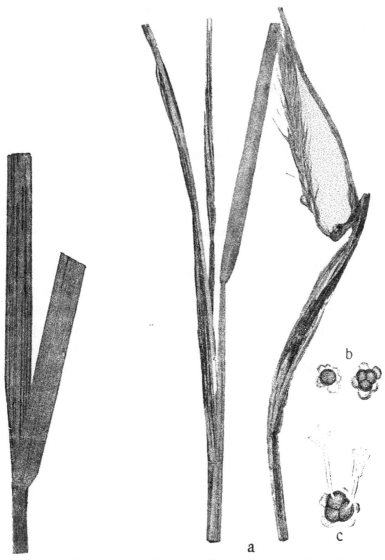

FIG. 39.—SMUT OF GIANT MEADOW-GRASS—*Ustilago longissima*. (THE AUTHOR.)

FIG. 40.—STALK SMUT OF RYE—*Urocystis occulta*. (*a*, THE AUTHOR.)

a, Smutty straws; *b*, spore-balls not germinated; *c*, a germinating spore-ball.

Stalk Smut. (*Urocystis*.)

The spores, numbering about ten, are joined into spore-balls or clusters, with one or more larger and darker spores in the centre and several smaller and lighter spores outside. At the germination there grows out from each one of the inside spores a basidium, having at the top a crown of spool-formed spores.

Stalk Smut of Rye. (*Urocystis occulta.*)

The spore masses originate on the stem and leaves of rye in long, parallel stripes, at first covered by the cuticle of the organ and of a lead-grey colour. Finally the epidermis breaks, and the black spore-stuff becomes visible. The diseased straws are more or less distorted and stunted in their growth. Usually there is no ear-formation, or else the ears are empty. Generally every straw in the plant is affected. In sporadic cases the disease has infected wheat.

The disease is not of a very frequent occurrence. Usually it is limited to sundry specimens in the field. In Denmark cases are known where it has caused a reduction in the crop of 50 per cent. At the threshing of such rye the air is filled with spore-dust to such an extent that the labourers are subject to nausea.

PROTECTIVE MEASURES.—The same as against stinking smut on wheat.

CHAPTER VI

RUSTS—UREDINACEÆ

THESE fungi are, like the moulds and smut - fungi, veritable parasites. They develop their mycelium as a mesh of fungoid threads inside the host plant between the chlorophyl cells. Inside the cells the mycelium, as a rule, sends only short side-branches, sucking warts, or suckers. Sometimes there is found, also inside the cells, a largely ramified mycelium. The cells are not at once destroyed by the fungus, but retain their natural shape for some time. Gradually they are, however, pressed together by the constantly growing mycelium, and all the parts of the cells—viz., their walls, chlorophyl bodies, and so forth—are overcome and consumed by the fungus. Finally there is formed a homogeneous *hymenium*, from which proceed long strings which bear spores. Then the cuticle of the host plant breaks, and an open wound appears, filled with a yellow or brown mass of dust. Sometimes the epidermis remains unbroken over the stuff like a transparent film.

The rust fungi have usually several forms of spores. In the summer there develop summer spores (*uredospores*). These are yellow, or yellowish-red, unicellular, and warty or prickly on the surface (see Fig. 42, *b* and *c*). They can, as a rule, germinate at once. The germinating filaments penetrate the stomata of the infected organ and bring on fresh sores within a week or two. During the course of the summer several such generations of spores may germinate.

From the same mycelium there will later on in the summer and in the autumn come forth autumn and winter spores (*teleutospores*).

63

These are brown or black, both unicellular and many-celled, with thicker walls (see Fig. 42, *e*). Occasionally they germinate at once, but in most cases they are hibernating spores, and do not grow before the following spring.

The germination of these spores takes place in such manner that from each compartment of the spore there grows out a short *basidium* (see Fig. 42, *f*), which is divided or jointed, and from each one of these joints a basidium spore (*sporidium*) is separated and carried away by the wind. Should this spore reach a growing leaf, or any other part of a plant that is susceptible to this fungus, then the spore will penetrate its epidermis by means of a germ-thread, and thus get access to the inside of the organ, later on becoming evident by rust sores on the surface.

This invasion may be either on the same host species where the spores have bred, or on a plant closely related thereto, or on a plant that is far apart from them. In the latter case the fungus is said to be " host-changing " (*heterœcious*), in the two former cases " non-host-changing " (*autœcious*).

That stage of the fungus which originates at its change of host plant, differs considerably from the previously described stages of summer spores and winter spores. It is called the " cluster-cup stage " (*Æcidium*, see Fig. 43), and embraces, as a rule, two different stages of development. One of these forms is called *spermogonia*, and consists of very small flask-like formations arranged in groups, which are sunk in the leaf. These little vessels contain a large number of thin rod-like *spermatia*. The other form, which often appears on the opposite side of the leaf, consists also of small groups of closely crowded cups, having reflexed or turned-out ragged edges (see Fig. 43, *b*), or else bare spore-heaps (*Cæoma*). The æcidiospores are arranged in bead-like or moniliform rows. These spores, as a rule, are able to germinate at once, and produce, within eight to ten days, sores with summer spores, if they strike upon a susceptible host plant or a corresponding uredo-carrying species.

However, all rust fungi do not possess all these kinds of spores; frequently one or more of them may be missing.

The most important of the spore-form are the teleutospores, because the fungus usually hibernates with them. From these spores the systematic arrangement of the rust fungi is made and the genera named. The different species are, however, called after the host plant.

Some of the rusts can live on several species of host plants that are related to each other (*heterofaguous*), but others are confined for their sustenance to only one species (*isofaguous*).

By the " specialising " of .the parasitismus is meant the fact that within the very same species of fungus can be discerned several different biological races, "specialised forms," in their appearance quite like one another, the differences being confined to their inner nature, and manifested in the circumstance that each of these forms lives on its own—either one or more—species of host plants.

In those cases where it has proved insufficient to explain by means of hibernating spores the reappearance of the disease the following year, it has been said to depend upon a hibernating mycelium resting in the stalk or in the rhizoma of the host plant, or also upon hibernating uredospores.

The truth of the former supposition is hardly gainsaid. But different opinions exist with regard to the hibernating uredo stage being a link in the life-history of the rust fungi. It is surely a fact that there are occasionally found, sometimes in one, sometimes in another, kind of rust, sundry hibernating uredospores, that are able to germinate. And as these are found towards the spring—in March, April, and May—it seems that a fresh outbreak of the disease during the summer might originate from them, but this is by no means proved. A critical survey of the known observations is opposed to such a supposition.

From all the extensive investigations, which have been carried on during the last fifteen years with regard to the nature and development of the rust fungi on seed, it appears as if in these fungi there should be distinguished—beside the well-known vegetative mycelium stage—another vegetative stage, when the fungus exists in the cells of the host plant as a formless plasma body, a sort of *plasmodium*, symbiotically fused with the protoplasm of the cells, and forming together with these a *mycoplasm*. The mycoplasm-

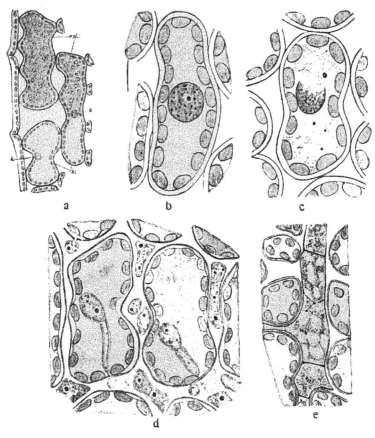

FIG. 41.—MYCOPLASM AND ITS TRANSFORMATION INTO A MYCELIUM.
(THE AUTHOR.)

a, Leaf-cells of a germinating plant of autumn wheat, forty-four days after sowing, partly containing mycoplasm; *mpl*, in dormant stage; *k*, nucleus; *kl*, chlorophyl corn; *b*, leaf-cell of autumn rye-plant, one to two weeks before the summer outbreak of the brown rust on rye, the diseased nucleus enlarged; *c*, leaf-cell of a similar plant, from the neighbourhood of the point where the first uredo sores broke forth, the mycoplasm being in the stage of maturing, the nucleus in a state of dissolution, and miniature nucleoli forming in the plasm; *d*, leaf-cells of oat-plant, from the neighbourhood of the point where the first uredo sores broke forth, the mycoplasm penetrating into the place between the cells, where it forms a mycelium; *e*, young stage of a mycelium with conspicuous separating walls.

carrying cell presents otherwise a normal appearance, with nucleus, chlorophyl bodies, and so forth. There cannot be recorded any parasitical fungoid life that would waste away the host plant. We may surmise that the fungus in this way can exist in most of the chlorophyl-carrying cells, up to the ears and bloom, in all sorts of seed that are especially suitable for the fungus, or, as it is expressed, are in a higher degree susceptible.

The period during which the fungus exists in this latent state varies in different cases. From four to five weeks, it might last for as many months and even for some years. This is the *dormant* stage of the mycoplasm.

Sooner or later, at a certain period of the life of the host plant, at a certain season, and with favourable environment or circumstances (soil, moisture, warmth, light, and so forth) for the development of the fungus, and varying with different sorts of rust, there will commence a new stage in the existence of the mycoplasm—the stage of *maturing*, when the fungus forces its way out from the symbiotic complex, penetrates the walls of the cell, and develops an intercellular mycelium. This maturing seems to be of short duration: it lasts only for a day or two, or possibly only some hours. As soon as the intercellular mycelium begins to form, it takes generally one week before open rust sores with spore-stuff begin to appear on the surface of the plant.[1]

Two-Celled Rust. (*Puccinia.*)

Among the genera of the rust fungi this one contains the greatest number of species. The winter spores are generally bilocular and contained in patches of sores, round or in stripes, either uncovered or occasionally covered by the epidermis of the host plant. Sometimes there may appear one-celled winter spores mingled with the double-celled.

[1] Those who wish to learn more about this new doctrine regarding the inherent character of the rust disease, and also about the foundation of this new theory, are referred to the following works upon the subject by the author of this book: " Über die Mycoplasma-Theorie, ihre Geschichte und ihren Tages-stand " (" Biol. Centralbl.," 1910, p. 618), and " Der Malvenrost, seine Verbreitung, Natur und Entwickelungsgeschichte " (Kgl. Sv. Vet.-Akad. Handlingar, vol. xlvii., No. 2, Stockholm, 1911).

GRAIN RUSTS.

(i.) SPECIES THAT SHIFT THEIR HOST PLANTS.

Black Rust of Cereals. (*Puccinia graminis.*)

This is the most noticed and best known of all the different species of grain rusts. It not only infests all our cereals—rye, wheat, barley, and oats—but also many grasses. It appears in Central Sweden about the middle or towards the end of July; a few weeks earlier on the autumn sown plants than on those sown in the spring. It is in the shape of longer or shorter reddish-brown dust-filled sores scattered on the leaves and stalks, especially on the sheaths, but also on ears and panicles. These sores are the stage of the summer spores of the fungus (*uredo*). The spores are long, covered with prickles, and supplied at the narrower middle part with two opposite germ pores. Exposed to moisture (rain or dew) they will germinate in a few hours. If this germination takes place on a suitable substance, say if spores from oat germinate on a young oat-leaf, then the germ-thread forces its way into this and produces on the maculated spot, within eight to ten days, a group of fresh sores, filled with similar spore-stuff. For the following two to three months generation will follow generation as long as there is any supply of fresh green sprouts.

Until a few years ago it was believed that spores from one kind of seed could convey the disease to another, and also that seed could get diseased from different sorts of grasses, and in its turn convey the disease over to them. This idea is, however, erroneous. Within the species of black rust there exist several specialised forms, more or less adapted to their own host plant or plants.

As the species appear in Sweden, there can be distinguished the following six forms:

(A) Not distinctly fixed (occasionally going over to other forms of grass): (1) f. sp. **Tritici** on wheat (seldom on rye, barley, and oat).

(B) Distinctly fixed (firmly confined to the indicated species): (2) f. sp. **Secalis** on rye, barley, and on couch grass (*Triticum repens*),

Elymus arenarius, Bromus secalinus, and others; (3) f. sp. **Avenæ** on oat and on *Avena elatior, Dactylis glomerata, Alopecurus pratensis, Milium effusum,* and others; (4) f. sp. **Poæ** on *Poa compressa* and *P. pratensis*

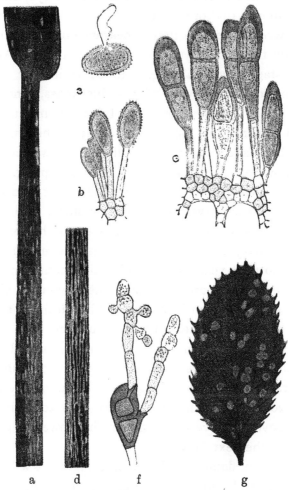

a   d   f   g

FIG. 42.—BLACK RUST OF OAT—*Puccinia graminis.* (THE AUTHOR.)

a, Stem of oat with uredo sores; *b* and *c,* summer spores—one of the spores germinating; *d,* straw of oat with puccinia sores; *e* and *f,* winter spores—one of the spores germinating; *g,* barberry-leaf with cluster-cup rust.

(5) f. sp. **Airæ** on *Aira cæspitosa* and *A. bottnica;* (6) f. sp. **Agrostis** on *Agrostis canina* and *A. stolonifera.*

An oat-plant infected with black rust can thus in its turn infect only oat, but not wheat, rye, barley, and so forth. In different countries, however, the specialising of a fungus species may take place in different ways. Thus, as an example, the black rust form of barley is in Sweden the same as that of rye, but in North America the same as that of wheat.

About two weeks after the first appearance of the summer spore stage, generally the end of July or the beginning of August, the winter spore stage (*Puccinia*) begins to show itself. This forms long streaks of sores, filled with a black dust-mass, hence the name "black rust." The winter spores are bilocular and thick-walled, narrowing off towards the top. After a natural hibernation in the open, where they have been exposed during the winter to air and wind, these spores are able to germinate the next spring. But if the spores during the winter have been protected, either by cold or warm surroundings, or if they have been in a grain stack, then they can by no means be induced to germinate. Rusty straw that has been laid either in the granary or in a stack can thus be considered as harmless in respect to next year's crop. The hibernating spores, which have been free in light and air, have their greatest germinating power in the spring, in April or May. This declines towards the summer, and has generally ceased by the autumn.

The power of germination is not exhausted through ploughing down the rusty straw and stubble deeply into the ground. Rusty portions of straw that have been dug down in the middle of October as deeply as 20 to 50 centimetres, or say 8 to 20 inches, have in experiments proved to have still retained the power of germinating when unearthed the next May.

Exposed to warm, damp air, the hibernating spore will germinate. Each of its cells sends out a germ-thread, a basidium, which divides itself into several (as a rule four) joints. From these joints basidium spores are borne, and these will, in their turn, germinate.

Should this germination take place on the young leaf, or sprout, or fruit of the barberry-shrub (*Berberis vulgaris*), or on the young fruit of the *Mahonia Aquifolium*, occasionally cultivated in our gardens, then there appears on these the barberry cluster-cup

(*Æcidium Berberidis*). On the barberry-leaves the rust spots are almost circular, on the upper side red, with numerous tiny black pricks, these being the openings of the flask-shaped spermogonia embedded in the leaf. On the under side they appear as large yellow patches, with numerous closely-massed open tube- or cup-like æcidia. In the spermogonia there develop small staff-formed spermatia, whose function is, as yet, not fully known. The æcidium tubes have a ragged, reflexed edge, and they contain numerous æcidium spores arranged in bead-like strings from the bottom.

The barberry cluster-cup appears earliest during the last days of

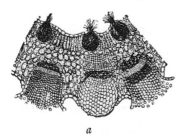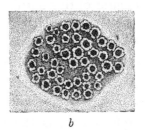

a   *b*

FIG. 43.—BARBERRY CLUSTER-CUP—*Æcidium Berberidis.* (*b*, THE AUTHOR.)
a, A barberry-leaf section, showing spermogonia above and æcidia below ; *b*, patch with cluster-cup from the under side of the leaf.

June, or the first days of July. The spores germinate somewhat irregularly and capriciously. The germination is promoted to some extent by chilly nights with intervening warm days. If spores from a cluster-cup spot, which has originated through contamination with a rusty oat-straw, should happen to germinate on a young oat-leaf, then there will arise in about eight to ten days a group of sores in the stage of the summer spores of the fungus. The same will happen if the germination has taken place on young leaves of some of the other grasses that are susceptible to the same form of black rust as the oat—viz., cock's-foot grass (*Dactylis glomerata*), meadow fox-tail (*Alopecurus pratensis*), and others. But should it happen that the germination of this cluster-cup form takes place on rye, wheat, barley, and so forth, then no rust sores will appear. In the same way cluster-cup, which originated through contamination from rusty rye- and barley-straws, will provoke rust sores on rye and barley, but not on wheat, oat, and so forth.

This difference of forms just described in the barberry cluster-cup, as to the existence of which we have only recently become aware, has naturally a great influence in regard to the rôle of the barberry-shrub as an agent in the spread of black rust to the grain-fields. A barberry-shrub that has been infected through rusty oat, cock's-foot grass, or meadow fox-tail can spread rust to an oat-field, but not to rye-, wheat-, or barley-fields. By this the share of the barberry-shrub is rather limited in the occasional devastations of cereals by black rust. This also explains why current opinions differ concerning the danger of this shrub.

Most persons who have recently devoted themselves to a special study of the rusts of cereals agree now upon this point, that the influence of the barberry-shrub is by no means so important as some time ago it was thought to be, this recent opinion being the result of fresh investigations and discoveries. One of these discoveries, already referred to, is the capricious nature of the spores of the cluster-cup in regard to germination. The cases where germination has succeeded are but few in comparison with those that have failed. Another observation is that the capacity to spread abates with the distance from the shrub   It remains yet to be proved that any contamination is incurred at a greater distance than 25 to 50 metres (say 75 to 150 feet) from the shrub. It is also now acknowledged that the distribution of black rust is in no way proportionate to the greater or lesser existence of the barberry-shrub in a district. On the contrary, it happens that black rust is very destructive in places where neither barberry nor any other carriers of this cluster-cup exist.[1]

Neither has it been proved that occasional legislative measures

[1] It must specially be pointed out that black rust is destructive to Australian whea to the extent of nearly £1,000,000 sterling per annum, although the barberry-shrub is not indigenous there, and the imported specimens never show signs of cluster-cup. (See "The Rusts of Australia," by D. McAlpine, Melbourne, 1906, pp. 64, 66 *et seq.*)

for the destruction of the barberry-shrub in one locality or another have prevented or reduced the destructiveness of the black rust.[1]

It is a matter of course that with regard to the black rust the same circumstances that affect other rust fungi, such as the level of the field, the drainage, the physical and chemical conditions of the soil, the fertilizers, the previous crop, the season of sowing, the state of the atmosphere, and so forth, have a material effect upon the more or less malignant nature of the disease. In spite of numerous observations and experiments for the purpose of solving all these interesting problems we are still at a loss to give definite advice to the planter. Only general observations can be furnished.

It has not been discovered whether different plants possess in any degree different powers of resistance against this fungus.

PROTECTIVE MEASURES.—See pp. 84, 85.

Brown Rust of Rye. (*Puccinia dispersa.*)

This rust only settles on rye. The time when the summer spores appear is generally the middle of June for rye sown in the autumn, and a few weeks later for rye sown in the spring. Occasionally there may be seen solitary uredo sores in the rye-fields in the autumn about a month after the sowing, and also very early in the spring when the snow has melted away. This proleptic outbreak of the disease terminates of its own accord sooner or later, and is in no direct connection with the principal outbreak in the summer.

The uredo sores form on the leaves, generally on the upper side, in small, chocolate-brown, irregularly scattered patches. The spores are globular, and, as a rule, germinate readily.

A few weeks after the outbreak of the uredo sores there appear on the under side of the leaves the black groups of the teleutospores, which are covered by the epidermis of the leaf. In this kind of rust these spores are able to germinate the same autumn, as soon as they have reached their full maturity. Should their germination take place on young leaves or other parts of *Anchusa arvensis*, or

[1] See " Die wahre Bedeutung der Berberitze für die Verbreitung des Getreide-rostes," von J. Eriksson, " Illust. Landw. Zeit.," Berlin, May 22, 1907, No. 41

A. officinalis, then the cluster-cup (*Æcidium Anchusæ*) develops on these.

This form of cluster-cup is found frequently in the southernmost part of Sweden and in Denmark during the months of August and

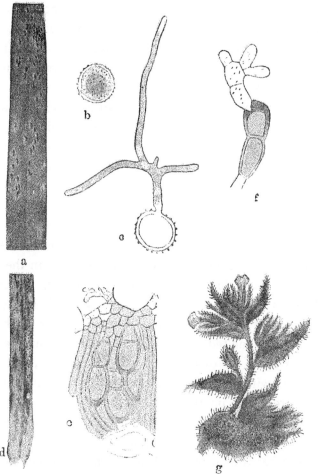

FIG. 44.—BROWN RUST OF RYE—*Puccinia dispersa*. (THE AUTHOR.)

a, Rye-leaf with uredo sores; *b* and *c*, summer spores—one of the spores germinating; *d*, rye-leaf with puccinia spots; *e*, autumn spore group; *f*, autumn spore germinating; *g*, branch of *Anchusa arvensis* with cluster-cup.

September on two weeds (*Anchusa arvensis* and *A. officinalis*) in potato-fields, on the edges of ditches, and so forth; but in Northern and Central Sweden it is rare. These two weeds are perfectly sound in these districts, even in places where rye, seriously affected with brown rust, grows in the vicinity.

The spores of the cluster-cup grow readily and quickly. If there are young sprouts or plants of rye close by, these will become contaminated, and uredo sores appear on them in eight to ten days.

With the approach of winter all the upper parts of the *Anchusa* species fade away—*Anchusa arvensis* dies altogether—only their seeds remain, and thus also perishes the cluster-cup stage of the rust, which can no more be an agent for a fresh outbreak of brown rust of rye during the following year. Neither does any hibernation of teleutospores take place, as these germinate during the autumn. Add to this that, in spite of careful research, no hibernating uredo stage has been found, either in the shape of spores or of mycelium. Hence the only explanation that remains is that the hibernation of this fungus takes place solely in the shape of a vegetative plasm.

PROTECTIVE MEASURES.—See pp. 84, 85.

Crown Rust of Oats. (*Puccinia coronifera.*)

This kind of rust affects oats amongst the cereals and also several forms of grasses. On oats the uredo stage appears by the end of July or the beginning of August. It forms shorter or longer orange-tinted sores on the leaves and the sheaths. The spores are globular, and germinate easily. A few weeks later the winter-spore stage sets in, as black spots placed in ring form around the uredo sores, covered by the epidermis of the leaves and sheaths. The teleutospores bear at the top a crown of blunt projections or processes, hence the name of "crown rust." They are genuine winter spores, and will not germinate before the following spring, after their natural hibernation.

In Northern and Central Sweden this kind of rust is only of subordinate consequence; it does not even occur every year. In more southerly countries, as in Germany, it is abundant, and seems there to take the place of oat rust instead of the black rust.

Should the germination of the winter spores, after their hibernation, take place on a young leaf of *Rhamnus cathartica*, then there will appear within eight to ten days a kind of cluster - cup (*Æcidium Catharticæ*). The spores of this cluster-cup germinate readily, and in their turn give origin to uredo sores on all the species of grass that are susceptible.

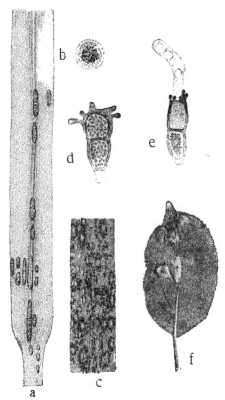

FIG. 45. — CROWN RUST OF OATS—
Puccinia coronifera. (THE AUTHOR.)

a, Oat-leaf with uredo sores; *b*, summer spore; *c*, oat-leaf with puccinia spots; *d* and *e*, winter spores—one of them germinating; *f*, leaf of *Rhamnus cathartica*, with cluster-cup.

This sort of rust appears on many other grasses, abundantly on rye-grass; but the forms on these grasses are individually adapted to them. Of this species we know the folowing specialised forms : (1) f. sp. **Avenæ** on oat. (2) f. sp. **Alopecuri** on *Alopecurus pratensis*, to which in sundry cases oat is susceptible. (3) f. sp. **Festucæ** on *Festuca elatior* and other kinds of *Festuca*. (4) f. sp. **Lolii** on English rye-grass (*Lolium perenne*); in some cases *Festuca elatior* is susceptible to this. (5) f. sp. **Glyceriæ** on *Glyceria aquatica*. (6) f. sp. **Agropyri** on *Triticum repens*. (7) f. sp. **Epigæi** on *Calamagrostis epigeios ;* in rare cases oat is susceptible to this form (8) f. sp. **Holci** on *Holcus lanatus*.

PROTECTIVE MEASURES.—See pp. 84, 85.

Another very closely related species of crown rust attacks several grasses ; this is **P. coronata,** which develops its cluster - cup on *Rhamnus Frangula,* and shows the following specialised forms : (1) f. sp. **Calamagrostis** on *Calamagrostis arundinacea,* and others ; in sporadic cases *Phalaris arundinacea* is susceptible. (2) f. sp. **Phalaridis** on *Phalaris arundinacea ;* only seldom *Calamagrostis arundinacea* is susceptible. (3) f. sp. **Agrostis** on *Agrostis vulgaris* and *A. stolonifera.*

Of the genus *Puccinia* the following grass rusts that shift their host plants may be noted :

FIG. 46.—PUCCINIA ARRHENATHERI. (THE AUTHOR.)

a, Uredo sores ; *b,* puccinia spots on leaves of *Avena elatior ; c,* barberry-branch, infected the previous year, when the now diseased leaf-clusters were budding ; *d,* fully-developed witch-broom on barberry.

P. Arrhenatheri occurs on *Avena elatior* as small, round, yellow uredo spots on the upper side of the leaves. On the under side of the leaves there develop sores of winter spores, although sparingly. The winter spores germinate in the spring.

If this germination takes place on the tiny buds that sprout on the young branches from the thorn axils, then the fungus grows into them, and after a year provokes rust-infected clusters of leaves on the sprout which came from the bud.   The incubation period of the fungus is thus a whole year.   On the diseased branch sound leaf-clusters may alternate with the infected ones, as the contamination has not touched all the buds.   The diseased leaf-clusters have all their leaves covered first with spermogonia, and later on with æcidia (*Æcidium graveolens*).   The leaves, as well as the branch, are shrivelled.   The fungus remains in this way year by year on the barberry-shrub, and finally a constantly increased and tangled-up " witch-broom " is created.   This kind of rust does not contaminate any grass except *Avena elatior*.

P. Poarum is parasitic on *Poa pratensis, P. compressa, P. alpina, P. trivialis,* and others.   It forms small, yellow, scattered uredo sores on the upper side of the leaves.   On the lower side appear the ring-formed groups of winter spores, covered by the epidermis of the leaf.   This fungus develops cluster-cups (*Æcidium Tussilaginis*) on colt's-foot (*Tussilago farfara*) in round, conspicuous, orange-yellow patches on the leaves and sheaths.

P. Phragmitis forms large, long-stretched, first brown, then black, sore patches on leaves and panicle stalks of *Phragmites communis*.   It brings on cluster-cup (*Æcidium rubellum*) on several kinds of sorrel, as *Rumex hydrolapathum, R. maritimus,* and others, and also on rhubarb.

P. Magnusiana forms very small sores on leaves of *Phragmites communis*, but develops cluster-cup on *Ranunculus repens* and *R. bulbosus*.

(ii.) Species which do not Shift their Host Plants.

Yellow Rust. (*Puccinia glumarum.*)

Besides the black rust, the yellow rust takes a first place as a destroyer of our grain. This fungus infests wheat, barley, and rye, and several grasses.

In Sweden, as well as all over Northern Europe, the yellow rust is most disastrous on wheat, especially winter wheat. Its attack on winter wheat begins generally in the middle of June, or towards the end of that month. The fungus makes an appearance on the plants that already have grown to a height of about 3 feet, and it usually attacks some of the middle leaves that show strong growth and are of a dark green colour. It forms, on the upper side small, lemon-tinted uredo sores, arranged in long stripes. Every day fresh sores break out in addition to the previous ones, until after a week or so the whole surface of the leaf is either wholly or in greater part covered with yellow sores. The fungus seems to work upwards along the plant, leaf after leaf. In more severe cases it also appears in the ears, on the inside of the awns, and under the husk of the tender grains. The spores are globular, and germinate capriciously.

A few weeks after the first appearance of the uredo sores the teleutospore stage of the fungus is in evidence, first on those leaves that are already diseased and on their sheaths. These groups of spores form, especially on the sheaths, long stripes of very tiny brown, later on black, pricks and are covered by the epidermis of the sheath. In severe cases they finally attack the inside of the awns and the wall of the grain. If the disease has entered the grains, these will, when mature, become more or less shrivelled, and are called " rust grains."

The teleutospores germinate in the late autumn, when the winter wheat is sown and quickens. It is supposed that the fungus then penetrates into the germ-sprout, where it remains in a latent condition for one or several months, at length breaking forth in the shape of open uredo sores on the leaves and stalks of the mature plant. It must, however, be admitted that neither anatomical nor experimental proofs have so far been forthcoming in this respect. Many reasons and numerous observations point to this explanation as highly probable, and possibly we shall not have to wait long for a scientific confirmation of this hypothesis.

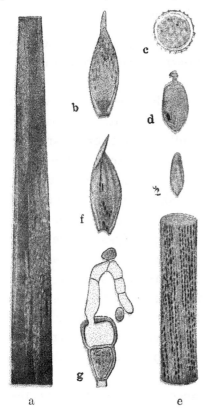

All experiments with a view to bringing on artificial contamination with this kind of teleutospores have so far been futile, on grasses as well as other plants. At any rate, it seems certain that no cluster-cup stage exists in Sweden.

As was the case with the previously described brown rust of rye, so with this fungus, it sometimes happens that, besides the normal and most prominent outbreak in the middle or end of June, other outbreaks, more or less severe, take place on the young leaves of winter wheat from the middle of October until the winter sets in, or otherwise in the early part of the following spring, say the close

Fig. 47.—YELLOW RUST OF WHEAT—
Puccinia glumarum. (THE AUTHOR.)

a, Wheat-leaf, and *b,* glume with uredo sores; *c,* summer spore; *d,* young wheat-seed with uredo spots beneath the husk, *e,* wheat-sheath, and *f,* glume with puccinia prickles; *g,* autumn spore germinating; *h,* shrivelled, mature rust grain.

of April or the beginning of May. None of these extra attacks are in direct connection with the principal attack during the summer. The autumn attack is broken off by the winter, and between the spring and the summer attack there exists, as a rule, a period free from rust. Both the autumn and spring attacks must be looked upon as " proleptic " outbreaks, which in some degree may be compared to those cases of plants of a higher order, that naturally bloom in the spring, but on account of special atmospheric conditions develop a number of flowers the previous autumn.

Even quite severe proleptic outbreaks need not be looked upon as certain indications of a coming severe rust year. The origin of this depends principally upon the climatic conditions during the months of May and June.

The following specialised forms of yellow rust are known : (1) f. sp. **Tritici** on wheat ; (2) f. sp. **Secalis** on rye ; (3) f. sp. **Hordei** on barley ; (4) f. sp. **Elymi** on *Elymus arenarius;* (5) f. sp. **Agropyri** on *Triticum repens* (couch-grass).

Different kinds of the same seed evince various degrees of susceptibility, and this especially refers to the winter wheats. We can distinguish between (a) *very susceptible,* (b) *less susceptible,* and (c) *almost insusceptible.*

FIG. 48.—YELLOW RUST OF WHEAT.
(THE AUTHOR.)

a, Tender wheat germ-sprout, with uredo sores on the first leaf (proleptic outbreak) ; *b,* a rusty leaf, cross-cut.

To the first of these groups belong a smaller number of wheats, among them Swedish white-eared velvet wheat, and several from North America, as Michigan Bronze, Horsford's Pearl, and Landreth's Hard Wheat. To the second group belong most of our ordinary wheats, containing those kinds which suffer but little in ordinary years, and only during " yellow-rust years " are

6

seriously affected. The last group, which never is seriously damaged, includes varieties of *Triticum durum*, *T. turgidum*, and *T. monococcum*, all forms of a more southerly nature, and consequently not easily adapted to North Europe. In the same group there have been placed several of the recently improved wheats. Experience has, after all, proved that the resistance of these sorts against the yellow rust is by no means consistent. Several cases are recorded where one kind, which for several years has shown great resistance, has changed its nature and been severely attacked. Neither theoretically nor practically can we consider the question of controlling yellow rust as yet satisfactorily solved by the improvement of cereals. Much remains to be done before we have fully conquered this powerful enemy.

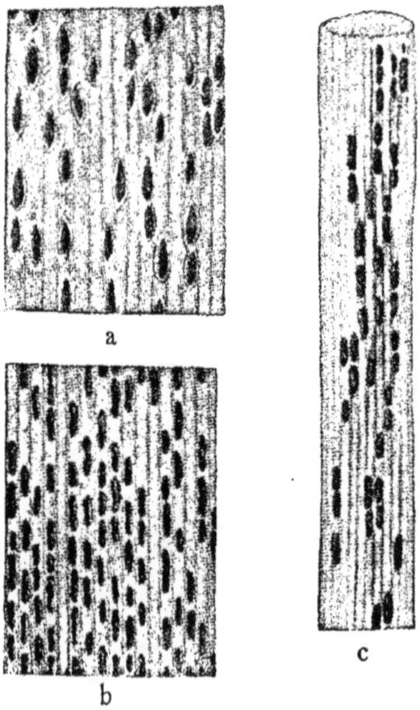

PROTECTIVE MEASURES.—See pp. 84, 85.

FIG. 49.—BROWN RUST OF WHEAT— *Puccinia triticina*. (THE AUTHOR.)

a, Wheat-leaf with uredo sores ; *b*, wheat-leaf, and *c*, wheat-straw with puccinia spots.

Brown Rust of Wheat.

(*Puccinia triticina*.)

This kind of rust attacks only wheat. It generally appears on wheat during the first or second week of June, and on spring wheat two or three weeks later. It forms small, scattered, chocolate-brown patches of uredo sores on the leaves, mostly on the upper side. The spores are globular, and germinate erratically, often feebly. A few weeks after the first appearance of these sores, if the attack be severe, the winter-spore stage of the fungus appears as black spore groups, arranged in stripes on the under side of the leaves and on the sheaths. These spore groups are covered with the epidermis of the plant. The spores, which look like their equivalents in the brown rust fungus of rye, hibernate, and attain their power to germinate in the following spring. All researches to detect any cluster-cup stage in this fungus have proved futile.

PROTECTIVE MEASURES.—See pp. 84, 85.

Dwarf Rust of Barley. (*Puccinia simplex.*)

This kind of rust attacks only barley. The actual time of its devastations sets in late in the summer, say the month of August.

It appears at first on the leaves as diminutive pale yellow uredo sores, eventually scattered over the entire upper surface. Later on it covers the under surface and the sheaths as brownish-black broken stripes. The winter spores are to a great extent unicellular. Sometimes the fungus also lodges on the ears and the setæ of the awns.

This form appears frequently as a destroyer in Denmark and Germany.

PROTECTIVE MEASURES.—See pp. 84, 85.

There may be mentioned also the following rusts of grasses, which do not either shift their host plants:

Timothy Rust (*Puccinia Phlei-pratensis*), that forms on leaves,

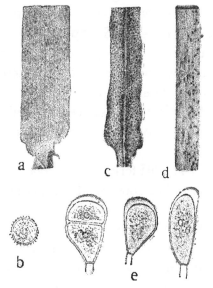

FIG. 50.—DWARF RUST OF BARLEY—*Puccinia simplex*. (THE AUTHOR.)

a, Barley-leaf with uredo sores; *b*, summer spore; *c*, leaf, and *d*, straw with puccinia spots; *e*, winter spores.

sheaths and straws of Timothy (*Phleum pratense*), and on *Festuca elatior*, brown, shorter or longer uredo sores, resembling the corresponding sores of black rust, and at one time generally considered as such. Occasionally there appear later on, especially on the straw parts, accumulations of black winter spores, resembling those of the black rust, but of more limited extent. Experiments to get this form to infect the barberry-shrub have proved unsuccessful, except in some sundry and doubtful cases. On other varieties of Timothy (*Phleum Boehmeri, P. Michelii, P. asperum*) outbreaks of genuine black rust (*P. graminis,* f. sp. *Avenæ*) have occasionally been observed.

P. bromina originates brown, and later on black, sore patches on several brome-grasses, as *Bromus arvensis, B. asper, B. secalinus, B. racemosus, B. mollis, B. sterilis, B. tectorum,* and others, but in Sweden, as far as we know, not on *B. erectus* and *B. inermis.* On these last two brome-grasses there have, in more southerly countries—for instance, Switzerland—been noticed a somewhat similar shifting sort of rust, called **P. Symphyti-Bromorum**, which develops the cluster-cup stage on *Symphytum officinale* and *Pulmonaria montana.* Often these two rust forms are considered synonymous, for what reason is not yet determined.

P. holcina causes reddish-yellow sore patches on the leaves, and later on, broken stripes on the sheaths of *Holcus lanatus* and *H. mollis.*

P. Triseti forms on *Trisetum flavescens* small, fairly evenly scattered, pale yellow sores on the upper side of the leaves, and later on, black spots on their under side.

P. Milii yields reddish-yellow, later black, sores on the leaves of *Milium effusum,* the wounds being usually surrounded by light rings.

P. Anthoxanthi appears as small, scattered yellow sores on the leaves of *Anthoxanthum odoratum.*

Protective Measures—(A) *Against Seed and Grass Rusts generally.* —(1) Avoid as far as possible moist, shady, and confined localities with poor drainage. (2) Cultivate the soil in such manner as experience has proved to be the best with regard to rapid growth and uniform maturity. (3) Avoid barn manure previous to sowing the spring seed, as such manure delays the ripening, and use among artificial fertilizers preferably the phosphorous ones, as these best promote the maturation. (4) When sowing, use a machine. (5) Sow early in the spring in a well-prepared soil, as suitable as possible. (6) Weed away from the vicinity of the grain-fields all such grasses as may possibly bring rusts to the seed; hence from the neighbourhood of rye- and barley-fields particularly *Triticum repens* (couch-grass), *T. caninum, Elymus arenarius,* and *Bromus secalinus;* and from the neighbourhood of oat-fields especially *Avena elatior, Dactylis glomerata, Alopecurus pratensis,* and *Milium effusum.*

(B) *Against the Rusts which Shift their Host Plants.*—(7) Remove the host plants that bear cluster-cups—viz., *Berberis vulgaris, Mahonia Aquifolium, Rhamnus cathartica, Rh. Frangula, Anchusa arvensis* and *A. officinalis*—to a distance of 25 to 50 metres, or 75 to 150 feet, from fields and pastures, from the edges of railways and other roads, from the yards around railroad stations, from cottage flower-patches and small orchards, from the unused ground of large orchards, and from the outskirts of woods. And do not introduce these shrubs into a vicinity where there are fields of seed susceptible to cluster-cup.

(C) *Against Yellow Rust.*—(8) Do not cultivate any sort of grain, nor a stock of such, if you are not quite satisfied that it never has been attacked by yellow rust—not even during a "yellow-rust year." (9) Do not imagine that the seed is immune, even if it has evinced no sign of disease during a "non-yellow-rust year." (10) Remember, experience has proved that specimens which once had great resistance alter their character and become more susceptible. (11) If a kind that previously has been endowed with great resistance begins to show signs of weakness, then avail yourself of the experience that has been gained in your country with regard to improvement in different sorts of seed, and choose for sowing some new variety of great endurance. (12) Rust grains shrivelled up by disease should not be used for sowing, for if a "yellow-rust year" should then happen to occur, it goes without saying that the crop fails. (13) When you choose your winter wheat, it is not sufficient to take into consideration its power only to resist the yellow rust, but also the resistance it possesses against cold winters and dry springs, its fertility and early maturity, and so forth.

Among other species of bilocular rusts that we have noticed may be mentioned **P. Cichorii,** which affects *Cichorium Intybus,* and forms small cinnamon-brown uredo sores on the leaves; but rarely are they followed by black puccinia spots on the stem. Another is **P. Spergulæ,** which produces brownish-black sores on spurry.

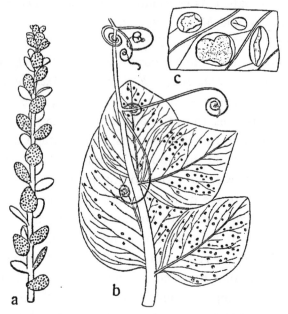

FIG. 51.—PEA RUST—*Uromyces Pisi-sativi.* (FROM P. DIETEL.)
a, Sprout of *Euphorbia Cyparissias* with cluster-cup ; *b* and *c,* pea-leaves with uredo sores.

Uromyces.

Rusts belonging to this genus have one-celled winter spores, usually egg-shaped, individually separated, and massed in small open spore groups.

(i.) Species Shifting their Host Plants.

Of these should be noted the following forms on grasses :

Uromyces Dactylidis, with very tiny, crowded, yellow, and later on black, sore spots on leaves and sheaths of *Dactylis glomerata.* The fungus has its cluster-cup stage on several species of Ranunculaceæ—viz., *Ranunculus polyanthemos, R. bulbosus, R. repens,* and *R. acer.*

U. Poæ causes similar sore spots on *Poa trivialis, P. nemoralis,* and *P. palustris,* and develops cluster-cup on *Ranunculus repens* and *R. bulbosus.*

U. pratensis is very similar to the previous one. It appears on *Poa pratensis,* and its cluster-cups come out in the early summer on *Ranunculus auricomus.*

Among forms belonging to this group are several rusts that attack leguminous plants :

U. Pisi-sativi brings on small, scattered, rusty-brown sore groups on leaves and stems of peas, and develops cluster-cups on *Euphorbia Cyparissias* and *E. Esula.*

U. striatus appears in a similar way on several sorts of clover, as *Trifolium agrarium, T. arvense,* and others, and on lucerne (*Medicago sativa, M. falcata, M. lupulina,* and others), and creates cluster-cups on *Euphorbia Cyparissias.*

U. Euphorbiæ-corniculati provokes similar sore groups on *Lotus corniculatus.* This sort has also a cluster-cup stage on *Euphorbia Cyparissias.*

(ii.) SPECIES THAT DO NOT SHIFT THEIR HOST PLANTS.

Beet Rust. (*Uromyces Betæ.*)

This disease appears sometimes in the spring on tender germ-sprouts of both sugar and fodder beets in distinctly limited yellow spots of cluster-cup. This same form appears occasionally also on those leaves which in the spring sprout out from planted seed beets. The disease, however, is most conspicuous at the height of the summer, when it forms numerous small brown patches scattered over the whole leaf on both its surfaces. These patches contain the summer spores of the fungus, and these can germinate at once and diffuse the infection over the beet-field. In the autumn the pale brown patches are replaced by dark brown, and these contain the winter spores of the fungus. The disease checks the power of assimilation within the leaves, and stops the growth of the root.

PROTECTIVE MEASURES.—(1) Pull off and destroy at once all the cluster-cup-bearing leaves that sprout from the seed beets. (2) Keep the field patch assigned for seed beets well apart from the regular beet-fields. (3) Take care that the beet seed used for sowing comes from a place where no disease has appeared.

To this group of rust fungi belong also several forms on leguminous plants:

U. Fabæ appears on *Vicia Faba, Pisum sativum, Lathyrus palustris,* and others. The cluster-cup of this form shows small rings on the under side of the leaves, but exists sparingly. However, during

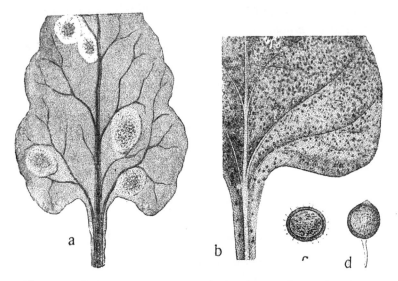

FIG. 52.—BEET RUST—*Uromyces Betæ.* (FROM O. KIRCHNER AND H. BOLTSHAUSER.)

a, Beet-leaf with cluster-cup ; *b*, beet-leaf with uredo and puccinia sores ; *c*, summer spore ; *d*, winter spore.

the height of the summer the cinnamon-brown uredo sores appear in great abundance ; later on hard, brownish-black, extended groups of winter spores appear.

U. Trifolii-repentis lodges on white clover. The cluster-cup stage breaks out on leaf-stalks and nerves, where it creates small distortions ; it also breaks out on the leaf-structure between the nerves, where it forms small rings. The heaps of brownish-black summer and winter spores appear later on, principally on the under side of the leaves.

U. Trifolii is very similar to the previous one, but has no cluster-cup stage. It appears on red clover and *Trifolium medium*.

U. Anthyllidis makes rusty-brown, and later on black, sore heaps on leaves and stalk of *Anthyllis Vulneraria*. With this is also placed a nearly similar form of rust fungus on lupin.

U. Onobrychidis creates brown, ultimately black, scale heaps on leaves and stem of *Onobrychis sativa*.

Melampsora.

The rust forms of this genus have unicellular winter spores, from cylindrical to prismatic, crowded side by side, and forming an agglomeration of brownish-black scales covered by the epidermis of the host plant.

To this kind belongs **Melampsora Lini,** which develops first yellow, and later reddish-brown to black, heaps of spores on leaves and stems of *Linum usitatissimum*. Under severe attacks the stems become frail and unsuitable for future use. Different sorts of flax have varying susceptibility. A rather similar form exists on *L. catharticum*.

CHAPTER VII

CLAVARIACEÆ

THE fruit bodies of these fungi are thread- or club-shaped, occasionally ramified like corals, of a light colour, and pulpy in their structure. Most of them thrive best in damp woods. The parasitic forms belong to the genus **Typhula**. Some develop their mycelium in living plants, kill these gradually, and develop forms on their surface, or their inside, called *sclerotia*. These are hard, tubercular masses, of the size of a cabbage- or clover-seed, seldom larger, and irregular in form. Their surface at first is white, and later on yellow, pink, or black, the inside pure white. The sclerotia separate finally from the host plant, remain unchanged during the winter, and germinate the next spring or summer. Hence they are the hibernating organs of the fungus. On germination the sclerotia extend one or several thin threads, either repeatedly ramified or single, but in that case expanded towards the top. At the thread-points basidium spores are borne, and these originate a new parasitical generation of the fungus and cause fresh outbreaks of disease.

To these belong the following forms:

Typhula graminum, at times noticed on tender plants of wheat and rye-grass. The mycelium penetrates and kills the plants. The sclerotia are of a reddish-brown colour, 1 to 2 millimetres across, and are rounded or edged. They develop on the lower straw and leaf. In the autumn of 1877 this disease appeared in a malignant form near Stockholm, on winter wheat from England (" Mainstay " wheat.)

T. Betæ appears on beets, mostly on the topmost parts of seed beets that are stored in cellars, but also on the stem and branches of growing seed-setting beets. The sclerotia are black, of the shape and size of a cabbage-seed. This fungus is frequently seen in Denmark.

T. gyrans exists on cabbage, turnip, and so forth, partly on the

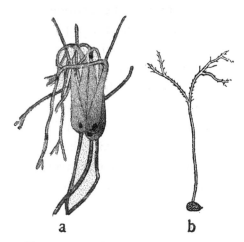 

Fig. 53. — Typhula graminum of Wheat-Sprout. (The Author.)

a, Withered wheat-plant with sclerotia; *b*, a sclerotium with fungus thread.

Fig. 54.—Typhula Trifolii. (From E. Rostrup.)

Two sclerotia with grown fungus threads.

leaves and their petioles, partly on the roots. The sclerotia are reddish-brown, turning black, and of the size of a cabbage-seed.

T. Trifolii attacks various sorts of clover. The points of the stalks of the host plant are then penetrated by a fine spawn, and shortly turn black. The sclerotia are brown, and finally black, being of the same size and appearance as clover-seed.

As a protection against these diseases, care should be taken in the first place that no sclerotia are mixed with seed used at sowing.

CHAPTER VIII

MILDEWS—ERYSIPHACEÆ

THIS group of fungi has, next to the rust fungi, the greatest number of species among the groups of parasitic fungi, and many of these species cause great devastation. They affect many dicotyledons of various families, but only certain grasses among monocotyledons. They appear abundantly on the leaves, especially on the upper side, and also often affect young sprouts, bloom, and fruit. They are provided with an air-spawn, which extends over the surface of the attacked host plant as a fine, white, cobweb-like mesh, or in a later stage as a dirty grey to black, thick coat. From this air-mycelium short threads (*haustoria*) extend into the epidermal cells below. By means of these the fungus derives its sustenance from the underlying plant tissue. In several cases there have been noticed mycelium threads also farther down in the host plant, partly in the space between the cells, and partly in the cells themselves.

Besides this, there are extended from the white mesh outwards numerous ramifications, which have at the top a shorter or longer chain of ovate bud-cells (*conidia*), that are liberated one by one. These give to the fungus covering a more or less flour-like appearance, hence the name of mildew. The conidia are easily scattered, either by means of solid objects, to which they stick, and with which they are removed from place to place, or by the wind, which carries them along for shorter distances. They germinate very readily. If this germination takes place on tender leaves or parts of sprouts belonging to a plant that is susceptible to this fungus, then mildew will appear in a week or so.

In the thick, felt-like cover, which usually forms later, will develop spore-cases (*perithecia*), first as yellow, clearly visible prickles, later brown to black. Seen through the microscope, they are either globular or oblate and altogether closed. The wall consists of a layer of small, crowded cells, some of which protrude long processes, simple or ramified. Inside the spore-cases are one or more egg-shaped spore-bags, which in their turn contain two, four, or eight colourless, unicellular spores. The spore-cases with their enclosed spore-bags remain unchanged over the winter. In the spring they break up and set the spores free. If the spores then germinate on a suitable plant, they will give rise to fresh outbreaks of disease.

In certain cases there is rarely any development of the spore-cases designed for hibernation, and this is prevalent from year to year over large territories, and even whole continents. In spite of this, the disease recurs every year. It is thought that the recurrence of the disease in such cases must be explained by the hibernation of the mycelium. It might also be possible that the fungus under one shape or another, which it is difficult to discover, can continue its life in the cell structure of the diseased plant.

The different species of mildew are characterised generally by the appearance of their "appendages," especially the ramification of these, and also by the number of spore-bags (*asci*) in each spore-case, and by the number of spores in each spore-bag. In the same species there is found—just as was the case with the rust fungi—several biologically separated "specialised forms," each one designed for its special host plant.

There are two reasons for the destruction caused by the mildew fungi: the air-mycelium keeps light and air from the plant, and the haustoria, which have penetrated the epidermis, drain it out. The affected parts wither and dry up prematurely. Those mildew fungi that trouble agricultural plants belong to the genus

Erysiphe.

The spore-cases contain several (eight to twenty) spore-bags, each spore-bag with two to eight, and sometimes twelve, spores. The appendages are very little, or not at all, ramified, generally like ordinary fungi threads.

Grass Mildew. (*Erysiphe graminis.*)

The air-mycelium of the fungus forms on the leaves and stalks in white or pink meshes, more or less extended, and of varying thickness. Some of the ramifications turn almost vertically outward and liberate bud-cells arranged like bead-strings; these germinate readily and diffuse the disease. This stage of development of the fungus appears early in the spring, shortly after the snow has melted, on fields where autumn seed has been sown. The fungus is often of a reddish-grey appearance, and is frequently taken for rust. If the weather be favourable for development, then it changes during the summer into a greyish-white thick felt, with numerous little brown or black spore-cases embedded in the felt.

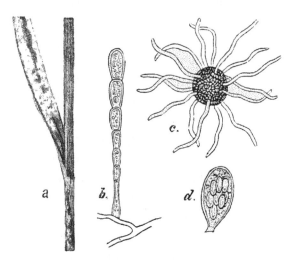

FIG. 55.—MILDEW OF WHEAT—*Erysiphe graminis.*
(*a*, THE AUTHOR ; *b, c, d,* FROM E. ROSTRUP.)

a, Wheat-straw with mildew felt ; *b*, chain of bud-cells ; *c*, spore-case with appendages ; *d*, spore-bag with eight spores.

Within a week or so after the spore-cases have begun to appear, the leaves that were covered by the mildew felt die, and at length the felt drops off. But the fungus hibernates in the spore-cases. In severe cases the disease may cause the premature ripening of the plants, when the grain becomes small and shrivelled.

Of this form of mildew there exist many specialised forms: (1) one on wheat, "German wheat," and other cultivated plants of the genus *Triticum ;* (2) one on cultivated barley, besides three special forms on other kinds of *Hordeum* ; (3) one on rye ; (4) one on oat and *Avena elatior ;* (5) one on several .of the genus *Poa ;* (6) one on *Triticum repens* (couch-grass) ; (7) one, and possibly four to five, on several species of *Bromus.* Whether this specialisation is carried out similarly in all places where mildew appears is not yet clearly known.

Occasionally the disease has proved to be very malignant on wheat. Such a case occurred in 1877 in California, when about 200,000 hectares (equal to about 500,000 acres) of wheat-fields were badly devastated, and also in the year 1885 near Stockholm and in the southern part of Sweden.

Among other forms of mildew that infest agricultural plants should be noted the following species :

E. polygoni (also called *E. communis, E. Martii, E. Pisi,* and so forth) forms on the leaves and stalks of many plants a thin; white, evenly

Fig. 56.—MILDEW OF CLOVER—*Erysiphe polygoni.* (THE AUTHOR.)

Leaf of meadow trefoil.

spread film, with numerous embedded black spore-cases—viz., on turnip, rape, peas, vetch, clover, lucerne, lathyrus, lupin, and so forth. Also in this species of mildew several specialised forms are concealed.

E. Cichoriacearum (also called *E. lamprocarpa, E. Linckii,* and so forth) attacks in the same way many composite plants (chicory, sunflower, aster), plants of the cucumber family, and others. Also here specialised forms occur—for instance, a special form on the cucumber family (at least seventeen different species, belonging to ten genera) ; another on the aster, and another on golden rod (*Solidago*), and so forth.

E. Heraclei causes a similar film on parsnip and other plants.

GENERAL PROTECTIVE MEASURES AGAINST MILDEW.—(1) Use only seed of plants free from mildew. (2) In case you think it will pay for the labour, plants covered with mildew should be powdered or sprinkled with some fungus-killing stuff—for instance, powdered sulphur, Bordeaux mixture (2 per cent.), or potassium sulphide solution (30 grammes of potassium sulphide—" liver of sulphur "—to 10 litres, or 2 gallons, water). (3) Do not allow stalks or leaves of plants (seed, peas, and so forth) that are abundantly covered with mildew spore-cases to be mixed in the manure or compost heaps.

CHAPTER IX

PERISPORIACEÆ

THESE fungi have a brown or black air-mycelium, which spreads itself all over the surface of the diseased plant. Occasionally it penetrates with ramifications into the plant. Outwardly it also

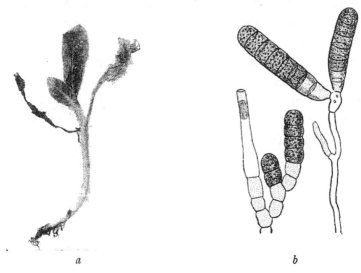

a   *b*

FIG. 57.—ROOT-ROT—*Thielavia basicola.* (*a*, THE AUTHOR ; *b*, FROM W. ZOPF)

a, A tobacco-plant, diseased ; *b*, fungus-mycelium with bud-cells.

sends out an ample supply of ramifications, which bear brown, unilocular bud-cells.

Root-Rot (*Thielavia basicola*) lives on the roots of peas, lupins, tobacco, and other plants. The roots shrink, become mellow and black, the stalks and leaves get stunted, and the bloom and fruit-setting ceases. The bud-cells have three to six loculi. The spore-cases are black and globular, with egg-shaped spore-bags. Each spore-bag contains eight brown, unilocular spores.

The disease was first noticed in England in 1850 on peas, and in Germany in 1876 on lupins, *Senecio elegans,* and other plants. In North America it was seen in 1891 on sweet violets, and in 1906 on tobacco. Since that time it has become a dangerous enemy of the tobacco-planter, not only in North America (from Ohio eastward), but also in Western Europe (between England and Italy). In Norway it has been malignant since 1910. In the tobacco planta-tions, which were severely attacked, the soil is alkaline, the fungus constantly being associated with this kind of earth.

PROTECTIVE MEASURES.—(1) Do not raise tobacco in contaminated . soil. (2) Keep the hotbeds well ventilated, and the field well drained. (3) Prepare the soil of the seed beds with a pint of commercial formalin mixed with $12\frac{1}{2}$ gallons of water, using $\frac{2}{3}$ gallon to each square foot. This should be done one week previous to sowing. (4) Do not use diseased seed-plants.

CHAPTER X

SPHÆRIACEÆ

In the fungi belonging to this group, which appear parasitic, it is usually only the conidia which develop on the live plant. These appear either through mycelium threads, rising freely from the surface of the plant, or else they are formed inside by means of pycnidia. The conidia will spread the disease from plant to plant during the vegetative period.

Later on spore-cases (*perithecia*) designed for hibernation develop on the plant. These are mostly sunk into the cell-tissue. They are small and globular, with a wart-like elongation, and provided with a narrow channel for the evacuation of the spores. From the bottom of the spore-case extend elongated spore-bags (*asci*), each one containing eight spores.

Leptosphæria.

The spore-cases at first are altogether covered by the epidermis of the host plant; subsequently they open and afford an exit for the spore-bags. The spores are elongated, light or brown yellow, and supplied with a larger or smaller number of traversing walls.

Straw-Breaker. (*Leptosphæria culmifraga.*)

This fungus, often also named *L. herpotrichoides*, attacks rye and wheat and is the principal factor in causing the straws to bend and finally snap at the base at the time of ripening. The fungus attaches itself to the lowest joint of the straw inside the sheath. The joint acquires a brownish colour, and sometimes there are seen on its surface a blackish cover or sundry small black groups of fungus threads. The cell-tissue of the straw is impregnated with fungus mycelium, and the mechanical elements of the tissue are weakly developed. The straw becomes weak at the affected spot,

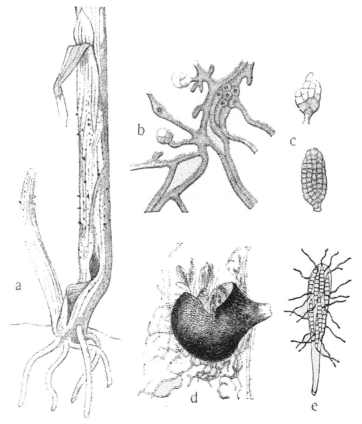

FIG. 58.—STRAW-BREAKER—*Leptosphæria culmifraga.* (*a* AND *d*, FROM F. KRUGER; *b*, *c*, AND *e*, FROM L. MANGIN.)

a, The lower part of a rye-straw with spore-case prickles; *b*, mycelium on the surface of the straw; *c*, conidia; *d*, spore-case pinched open; *e*, spore-bag with spores that have germinated.

breaking readily through the wind, or any other exterior cause. Those which are first broken do not develop ears, while others have weak ears, containing shrunken seed.

The conidia are elongated and many-celled. The spore-cases have an extended opening, and are often covered with aristæ. The spores are long, pointed, with numerous (three to eight) traversing walls. Frequently they will germinate inside the spore-bags.

A moist and cool atmosphere during the months of May and June seems to promote the course of the disease. The development of the plants is reduced, and their power of resistance against the attack of the fungus is diminished. Hence the disease is worst on cold and stiff soil.

This disease has been known in France since the middle of the last century, but general attention was not given to it there before the year 1870, and later on, when it was noticed on wheat. In the year 1894 it devastated the rye-fields in Northern Germany, reducing the crop even to the extent of 75 per cent. In Sweden it appeared in the year 1903, and has since reduced the wheat crop considerably.

Different sorts of wheat suffer in a different way. In Sweden the old rural kinds and their offspring have suffered worst, as, for instance, Upswedish hairy and Upswedish smooth rural wheat; while several new sorts, as Extra square-head, Fyris, Grenadier, Bore, and Top square-head wheats have been comparatively exempt.

PROTECTIVE MEASURES.—(1) Plough down the stubble early and deeply. (2) Do not omit to harrow in the spring the fields of winter seed. (3) Use a fertilizing of chili saltpetre for seed sorts that have shown only little resistance. (4) Select for cultivation those kinds that possess resistance and are otherwise adaptable. (5) Do not repeatedly sow the same sort of seed in the same field.

Black Pricks of Corn-Straw. (*Leptosphæria Tritici.*)

This fungus troubles all of our cereals. It was first observed in Italy on wheat, and has thence appeared in Germany on wheat, barley, and oat, in Denmark on barley and rye, and in Sweden on barley. It appears to the unaided vision as small, black prickles on the leaves, most conspicuous in the transparent sheath if this be held against the light. These prickles are globular spore-cases, having numerous spore-bags, each one containing eight four-celled spores. At the same time there are on leaves and sheaths in equal abundance lighter, brown prickles, containing staff-like, jointed bud-cells. These are thought to be an earlier stage of the development of the fungus.[1]

The disease can, in severe cases, considerably reduce the crop. The ears ripen prematurely. The corns become small and light.

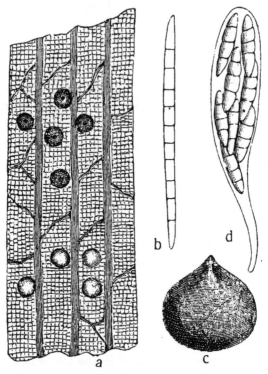

FIG. 59.—BLACK PRICKS OF CORN-STRAW—*Leptosphæria Tritici.*
(FROM E..ROSTRUP.)

a, Sheath with light pycnidia and dark spore-cases ; *b*, a conidium ; *c*, spore-case ; *d*, spore-bag with eight spores.

In the year 1895 this disease was ravaging the barley-fields in the vicinity of Copenhagen, and the loss was estimated at a million and a half Danish crowns, equal to over £80,000 sterling. Rainy weather in connection with a high temperature and but little wind are favourable for the progress of the disease.

[1] Another form of prick-disease, Grass Leaf-Spot—*Septoria graminum*, will be spoken of later on in this book, and likewise a third kind of prick-disease caused by the fungi *Sphærella exitialis* and *S. basicola*.

Black Spot of Cabbage.

(Leptosphæria Napi ; Sporidesmium exitiosum.)

This fungus troubles several plants of the cabbage tribe, as rape, turnip, white cabbage, cauliflower, and others. It attaches itself preferably to parts of the stem, bloom stalks, and sheaths, but now and again to the leaves. The fungus creates on the attacked parts either oblong-ovate or round blackish-brown spots. From these spots fungus threads proceed, and each one bears one conidium. These are elliptical, brown, provided with numerous division walls, and are drawn out to a point. In this stage of development the fungus generally is called *Sporidesmium exitiosum.*

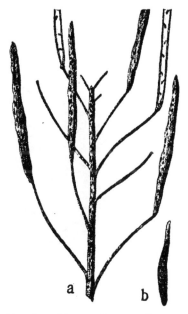

On the dead plant arise spore-cases which are thought to be a continuation of the fungus. The spore-cases are globular, with a short, extended mouth. The spore-bags contain long, yellow spores, with numerous traversing walls.

The disease causes damage especially to siliques. These become bent and brittle and open prematurely, thus preventing the full development of the seed. For prevention it is advised to reap early and gather the crop together for subsequent ripening.

FIG. 60.—BLACK SPOT OF RAPE— *Sporidesmium exitiosum.* (FROM O. KIRCHNER AND H. BOLTSHAUSER.)

a, Fruit cluster with black fungus spots ; *b,* conidium.

Pleospora.

Fungi of this group differ from those just described, principally from the spores being divided by longitudinal as well as transverse walls.

Sooty Mould of Beet.

(Pleospora putrefaciens ; Sporidesmium putrefaciens.)

The bud-cell stage of this fungus is called *Sporidesmium putrefaciens* (*Clasterosporium*), and at this stage it affects several sorts of beet.

The older leaves get yellow, and later on brown, and dry up.

FIG. 61.—BLACK SPOT OF TURNIP-LEAF—*Sporidesmium exitiosum.*
(THE AUTHOR.)

Finally, they are covered by an olive-brown, later on black, veil. This consists of fungus threads that have penetrated the epidermis from the inside of the leaf, and now are bearing brown, multi-locular bud-cells. In the withered leaves little black prickles (spore-cases) are forming, and ripen during the winter. In the spore-cases are confined spore-bags, each of which contains eight many-celled spores. If the weather be damp, then the diseased spots on the leaves become putrefied. Then it frequently happens that the whole leaf is spoiled.

As a preventive method, it is advised to speedily gather and burn all the infected leaves.

Stripe-Disease of Barley.

(Pleospora trichostoma f. Hordei erecti ; Helminthosporium gramineum.)

This disease becomes conspicuous early in the summer, towards the end of May or the beginning of June. It is especially six-rowed barley (*Hordeum vulgare*) and erect two-rowed barley (*Hordeum distichum erectum*) that are affected. The disease makes itself known by one or more long, narrow stripes on the blade. These stripes are at first pale yellow, then black, and edged with a narrow yellow

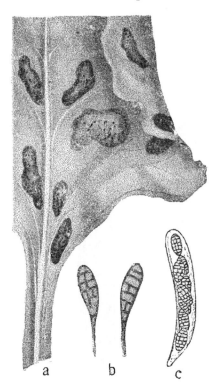

FIG. 62. — SOOTY MOULD OF BEET— *Sporidesmium putrefaciens.* (*a* AND *b*, FROM O. KIRCHNER AND H. BOLTSHAUSER; *c*, FROM A. B. FRANK.)

a, Leaf with fungus spots; *b*, bud-cells; *c*, spore-bag with eight spores.

brim. The number of the stripes is gradually increased, and finally the greater part of the blade is covered. At the same time the leaves begin to show a tendency to become puffed up in long threads, especially towards the top. The whole plant ceases to grow, and does not produce ears; or if these should appear, they will be found empty. Such diseased plants are scattered in various parts of the barley-fields. But they do not form any large patches, nor do they occupy a whole field.

On the stripes of the diseased leaves there soon appears a greyish-black stuff, and before long the whole plant is covered in the same way. This matter consists of short fungus threads, which bear

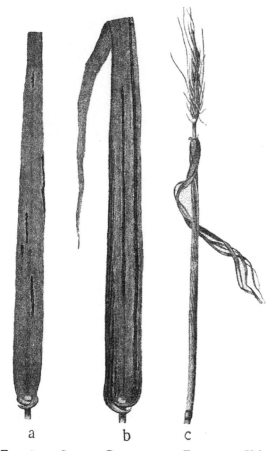

conidia and which push forward, one or many, through the cleavings of the leaf. These are roller-formed, and pro-vided with several—usually four or five—traversing walls. First they are yellow, then they become olive-coloured or brownish. They are attached very loosely, and become readily a prey to the wind. Some of them settle in the blooms of the barley-ear, thus in-fecting next year's seed. Plants raised from barley-corns developed from such infected bloom will in their turn get the stripe-disease.

a   b   c

FIG. 63. — STRIPE - DISEASE OF BARLEY — *Hel-minthosporium gramineum*. (*a* AND *b*, THE AUTHOR ; *c*, FROM *F.* KOLPIN RAVN.)

a, Early, and *b*, later stage of the disease on leaves ; *c*, top of a diseased straw.

It has been possible to follow the mycelium up to the vegetative point. The fungus ap-pears to spread in the same way as several kinds of smut—*i.e.*, through germ-infection. It reminds one also of the smut, inasmuch as the conidia have not the power to spread the disease during the summer to other leaves and plants.

This previously described parasitic stage of the fungus is generally named *Helminthosporium gramineum.*

This stage is followed by another, which lives saprophytically on the dead straw. This form has been noticed on the stubble of unhealthy plants. It appears as small, semi-globular sclerotia. These are black outside, and covered with stiff, dark, violet-hued aristæ. The sclerotia are like those of the following fungus—viz., " Spot-Disease of Barley "—only a little smaller. They remain un-changed during winter-time; but towards the spring they display renewed activity. This takes place either in such manner that conidia are yielded from their surface, or otherwise in such way that the conidia are changed into spore-cases. In the latter case the three or four exterior layers of cells of the sclerotia will form a wall of spore-cases, while the interior part is transformed and used for the creation of spore-bags. These each contain eight spores. By means of

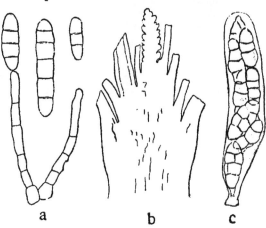

a   b   c

F ig. 64.—S tripe - D isease of B arley. (*a*, T he A uthor; *b*, from *F*. K ölpin R avn ; *c*, from F. N oack.)

a, Mycelium with conidia ; *b*, section of stem-top, with mycelium threads visible inside as black marks; *c*, spore-bag with eight spores.

these conidia and spores the disease can be communicated to very young germ-plants.

This disease is widely diffused through Europe. It affects almost solely six-rowed and erect two-rowed barley, and may in both these kinds reduce the crop by 20 to 30 per cent. Very early sowing and a low temperature at the time promote the development of the fungus.

P rotective M easures.—(1) Take seed for sowing from a field where the disease was absent the previous year. If you cannot entirely depend upon the immunity of the seed, then the following precau-tion should be observed : (2) Prepare the sowing-seed by dipping it about twenty times in warm water of 56° to 57° C., or for a period of, say, five minutes, and let it cool in the air. (3) Do not sow too early or when the atmosphere is chilly.

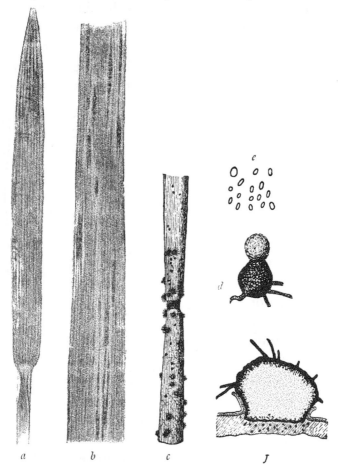

FIG. 65.—SPOT-DISEASE OF BARLEY—*Helminthosporium teres.*
(FROM F. KOLPIN RAVN.)

a, Young leaf in the primary stage of disease ; *b*, older leaf in the secondary stage of disease ; *c*, straw with pycnidia (small, black prickles) and sclerotia (larger accumulations with arista) ; *d*, pycnidia ; *e*, conidia ; *f*, a section of a sclerotium.

Spot-Disease of Barley.

(*Pleospora trichostoma f. Hordei nutantis ; Helminthosporium teres.*)

This disease—also called *Helmintosporios of oat*—appears on two-rowed leaning barley (*Hordeum distichum nutans*), and differs from the previously described stripe-disease in that it causes shorter, separate brown spots on the blade, while the leaf does not become swollen, nor does it lose its elasticity in the bottom part, but retains its usual form and position. The ear also attains more or less its normal development, and becomes only so far damaged that the points of the corns become brown.

The disease appears generally on all or most of the plants simultaneously, in large patches or over entire fields, and might attain the character of an epidemic at any time during the period of vegetation. This, however, is contingent upon the condition of the weather. Dampness promotes its progress.

There is a difference between primary and secondary spot-disease. The primary breaks out early, and only on the first leaf. It is reckoned that it originates from a source of disease in the seed that was sown. The secondary spot-disease appears on the later leaves. As no mycelium has been discovered in this case in the top of the stem —as in the stripe-disease—the conclusion is that the secondary stage is caused through infection from the primary stage.

a   b

FIG. 66. — HELMINTHO-SPORIUM AVENÆ OF OATS. (FROM F. KÖLPIN RAVN.)

The diseased parts of the leaves get covered by a greyish-black mass, consisting partly of fungus threads, partly of conidia, which are like those of the fungus of the stripe-disease. By means of the conidia the disease can travel from leaf to leaf during the vegetative period.

This form of development of the fungus has been called *Helminthosporium teres*. After this form there follows another, which saprophytically obtains its nourishment from the withered remains of the straw. This appears either in the shape of pycnidia, that are very tiny warty prickles, secreting small, round conidia, or it takes the form of larger sclerotia covered with aristæ, and these hibernate and attain a further development in the spring.

The primary spot-disease is prevented by the warm-water treatment; but with the secondary stage this is futile.

Closely related to these is **Helminthosporium Avenæ,** which attacks young oat-leaves. The spots are of a dirty grey or greyish-brown colour, rather frequently with a tint of red. As yet no late stage of this fungus has been discovered.

In Germany a similar spotted disease has been seen—**Pleospora trichostoma f. Bromi,** on *Bromus asper* and *B. inermis.*

Dry Spot of Potato-Leaf.   (*Sporidesmium Solani varians.*)

During the summer when the potato-plants are at their best, there appear on the leaves—generally first on the top leaflets, then on the side leaflets—more or less numerous, sharply-defined, blackish‑brown, desiccative blotches, frequently concentrically plaited. In severe cases the leaves finally dry up altogether and become black, and the plants die prematurely. In some places the disease has been so malignant that not a single potato-plant, not even one leaf, has been exempt. The tubers of the severely affected plants become small and deficient in starch.

The disease is brought on by a fungus called *Sporidesmium Solani varians.* On the older blotches appear small bunches of brown fungus threads, these being ramifications of a mycelium that lives in the leaves and in the hair formations. This mycelium can develop several sorts of spores. Some of these are large and multicellular, others small and mostly unicellular—*Cladosporium* (see further, p. 113). Occasionally pycnidia develop later in the autumn embedded in the leaf-tissue, and intended for the hibernation of the fungus.

The disease mostly afflicts early food-potatoes with thin, tiny leaves; not so much the late, thick-leaved potatoes.

It has appeared in Austria, Hungary, Germany, Denmark, and Sweden, being especially severe of late years in Bohemia and Moravia.

Presumably it is the same spot-disease of the leaves that is known in America under the name of " Early potato blight " (*Macrosporium Solani*), and in Germany by the name " Dürrfleckenkrankheit " (*Alternaria Solani*).

PROTECTIVE MEASURES.—(1) Gather up and destroy the diseased leaves. (2) As soon as the dry spots begin to appear, fungicides

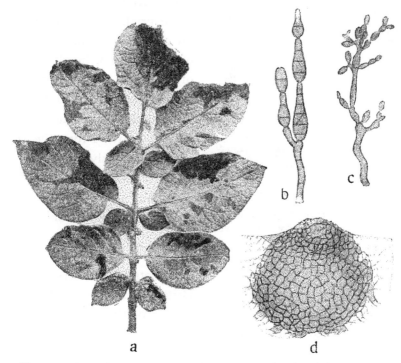

FIG. 67.—DRY SPOT OF POTATO-LEAF—*Sporidesmium Solani varians.*
(FROM J. J. VANHA.)

a, Diseased potato-leaf ; *b*, fungus thread with multicellular spores ; *c*, fungus thread developing conidia ; *d*, pycnidium embedded in the leaf-tissue.

should be resorted to. Sprinkle in July or August, when the leaves are first attacked, with 1 per cent. Bordeaux mixture, and repeat the treatment two to three weeks later on should it prove to be necessary. (3) Do not use potatoes for seed from a diseased field.

Somewhat similar fungi are the following : **Macrosporium Dauci,** which causes blackish-grey blotches on carrot-leaves. The leaves turn up and become dry. **M. sarcinæforme** on clover-leaves. **Alternaria tenuis** on tiny tobacco-plants.

Potato Dry Scab. (*Spondylocladium atrovirens.*)

This appears during the winter on the potato skin as irregular, white-spangled or pale violet blotches richly sprinkled with small

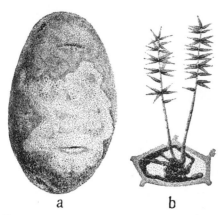

black prickles. Each one of these prickles displays a sclerotical mycelium on the surface of the potato or inside the cells of the skin. Pieces of the skin with sclerotia get into the earth, and remain there until the following year. From the sclerotia—previously known by the name *Phellomyces sclerotiophorus* —threads grow out, bearing several crowns of conidia, one above another.

a   b

FIG. 68.—POTATO DRY SCAB—*Spondylocladium atrovirens*. (FROM O. APPEL AND R. LAMBERT.)

a, Diseased potato; *b*, skin-cell with sclerotium, and from it spring fungus threads, producing conidia.

The disease has been reported in Austria, Germany, Great Britain, and America. It is not, as yet, fully proved that the sequence of this disease is a kind of potato-rot called "Phellomyces-Fäule."

Wheat-Ear Fungus. (*Dilophia graminis.*)

This very variable fungus afflicts a great many plants belonging to the family of grasses. It appears either with free conidia or with sunken conidia; or otherwise with spore-cases, containing spore-bags and spores.

In the free conidia stage it is found on the leaves, where it forms long, reddish-brown blotches edged by a light brim. The centre of the blotch is perfectly white, and carries spool-formed, colourless conidia, provided with three traversing walls, and usually with a few fine aristæ at both ends. This form of bud-cells (*Mastigosporium album*) appears on fox-tail grass (*Alopecurus pratensis*), Timothy-grass (*Phleum pratense*), cock's-foot grass (*Dactylis glomerata*), and other grasses.

These kinds of bud-cells, which are sunk in pycnidia, form greyish-black, white crusts. The bud-cells are long, and supplied with a bunch of fine aristæ. This form also produces distortions of the leaves in soft grass (*Holcus lanatus*), fox-tail grass (*Alopecurus pratensis*), and other grasses.

Occasionally it has been very malignant on winter crops in England, France, and Switzerland; and has to some degree caused the ears to become thin, black, and distorted, with only a few seed-corns in each ear ("Feder-buschsporen-Krankheit").

The spore-case form is very seldom developed in this fungus. The spores are long, narrow, many-jointed, and extended at both ends into a long point.

Ophiobolus.

Each spore-bag contains eight very long, thread-like, parallel spores.

"Take - All " and " White - Heads " in Cereals. (*Ophiobolus graminis* and *O. herpotrichoides*.)

Both these fungi, sometimes one, sometimes the other, appear on wheat, rye, and barley, frequently in company with the previously described straw-breaker (*Leptosphæria culmifraga*). Like that disease, these fungi also produce a black cover on the lowest joint of the straw, often giving the whole of the lower part of the plant a blackish hue. Also the

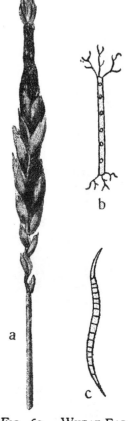

Fig. 69.— Wheat-Ear Fungus — *Dilophia graminis.* (*a*, From O. Kircher and H. Boltshauser; *b* and *c*, from E. Rostrup.)

a, Wheat-ear, distorted; *b*, conidium ; *c*, spore.

8

roots get affected and turn black for most of their length. If the plants are pulled up, then the earth frequently sticks to the roots. Some of the plants grow very scantily, with dwarfed ears. Others grow a little higher, but soon wither, and are black with blight up to the ears. The seed - corns become small and shrivelled.

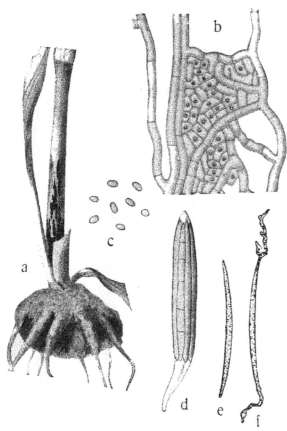

This disease is also called in England "Straw Blight," "Foot Rot," or "Black Leg," in France "Maladie du Pied," or "Piétin du Blé," and appears usually on limited patches in the field, not, like the straw-breaker, almost equally over the whole field.

The bud-cell stage (*Coniosporium*) forms a blackish dust on the straw below. The conidia are small, ovate, and uni-cellular. The spore-cases, which appear

FIG. 70. -- STRAW-BLIGHT —*Ophiobolus graminis.* (*a* AND *d*, FROM A. B. FRANK; *b, c, e* AND *f*, FROM L. MAGNIN.)

a, The lowest part of a prematurely-ripe wheat-plant; black fungus stripes on the base of the straw; earth adhering to the roots. *b*, mycelium ; *c*, conidia; *d*, spore-bag ; *e*, spore; *f*, spore germinating.

later in the autumn, are spadicose with wry neck embedded. In *O. graminis* they are smooth, in *O. herpotrichoides* hairy. There are eight spores, thread-like, colourless, narrowest and longest in the latter species. They show a row of small drops, and finally a few traversing walls.

O. graminis has proved very devastating on several sorts of wheat (Bore and Grenadier Wheat) in Sweden. *O. herpotrichoides* is especially known in Italy, France, and Germany. In Italy a similar disease has been noticed on Couch-grass (*Triticum repens*), and in Australia on Brome-grass (*Bromus sterilis*).

As a prevention against these fungi, it has been advised to burn the diseased stubble. For fertilizing, Thomas phosphate should preferably be used, whilst avoiding too rich nitrogenous manure.

Blight of Cereals.

(*Sphærella Tulasnii ; Cladosporium herbarum.*)

During wet seasons, towards crop time, there will be frequently found on rye, wheat, and occasionally other cereals, a black covering, on the yellow leaves, on the stem and the awns, and even on the seed-corns, which, seen through a magnifier, appears as small black dots, arranged in long stripes. Every one of these little dots consists of a bunch of fungus threads, olive-coloured, and at their points bearing simple or jointed conidia. The mycelium, from which this thread-bunch emanates, is spread either over the epidermis of the organ, or in its cells, or otherwise internally. It consists of an irregularly formed cluster of small, nearly black, tightly packed fungus cells, forming a sclerotium, occasionally acquiring a somewhat globular shape ; a sort of rudimentary spore-case.

FIG. 71.—BLIGHT OF WHEAT —*Cladosporium herbarum* (the conidium stage). (FROM E. JANCZEWSKI.)

a, Sheath, and *b*, spiculæ, with small, black fungus-dots.

In this previously described stage of development the fungus has from early times been named *Cladosporium herbarum*. It appears as if this form of the fungus has power to attack only such parts of a plant as have been enfeebled through some other agency. But it has no effect upon quite sound parts of a plant, unless the fungus, through artificial nourishment, has been given increased vitality. And through artificial nourishment it has also been possible to induce the fungus to further development. It has thus been noticed that globular sclerotia have been developed into actual spore-cases, with spore-bags.

Plants that have been severely attacked by this fungus yield small and shrivelled corns ; and in many places it has been noticed

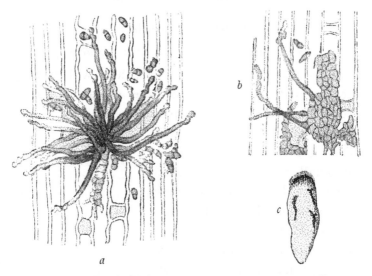

FIG. 72.—BLIGHT OF RYE—*Cladosporium herbarum.* (*a* AND *b*, FROM E. JANC-
ZEWSKI ; *c*, THE AUTHOR.)

a, Leaf-sheath with bunch of fungus threads, bearing conidia ; *b*, leaf-sheath with sclerotic fungus-heap, bearing conidia ; *c*, corn of "giddy-rye," with sclerotic formations on the surface.

that blighted seed, especially rye, that has been used in one form or another as food for man and beast has had a noticeable bad effect upon their health, although but temporarily. Headache, dizziness, exhaustion, and shiverings, often result from this poisoning. Such rye is in the rural districts of Sweden known by the name of " Örråg " = "giddy-rye." Most frequently prevalent and noted is such rye in Southern Russia ("Taumelgetreide"). But there, as in other places, other forms of fungi also appear on the poisonous seed-corns, especially the species described later on in this book, as

Fusarium avenaceum. This causes uncertainty as to which one of these fungi introduces the poisonous qualities to the seed. It is also to be noted that experiments with fodder for various animals, with seed infested with different sorts of these fungi, have given contradictory results.

Cladosporium herbarum lodges also on peas. The leaves get yellow or brown spots, on which are black bunches of fungus threads bearing

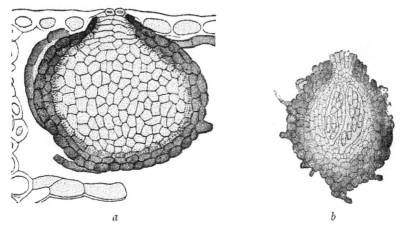

a   *b*

FIG. 73.—BLIGHT OF RYE (*a*), AND BLIGHT OF WHEAT (*b*)—*Sphærella Tulasnii* (the spore-case stage). (FROM E. JANCZEWSKI.)
a, Sclerotium beginning to develop into a spore-case; *b*, ready-formed spore-case, with two visible spore-bags.

bud-cells. The plants wither away from the root upwards and the blooming is feeble. In older plants the pods get affected.

In connection with this may also be described the following fungus on account of its great similarity in appearance, and the stage of conidium, although as yet no spore-case stage has been found:

Grey Spot of Oats. (*Scolecotrichum graminis f. Avenæ.*)

This disease appears in such a way that at the beginning of summer the leaves become spotted, and sometimes wither away before the formation of panicles, and this occurs on large or small patches in the fields. The plants either die, or they get so weakened that they only develop distorted panicles and empty withered spiculæ. The spots, which are large and brownish-grey, are at first few and limited, but extend more and more until the entire blade is almost or quite dead. On the grey spots are small blackish dots formed by bundles of fungus threads. These come from the openings, and consist of zig-zag formations of erect dull-brown fungus threads, from which proceed simple or cross-jointed conidia, having the same dull-brown colour.

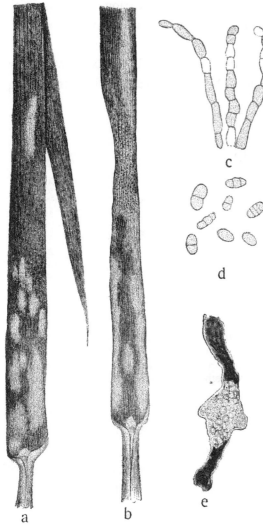

This disease returns usually on the very same patches every time oats are planted in the field. Occasionally the same phenomenon occurs, if spring wheat be sown in that field, while barley and winter crop are immune. The disease appears mostly on peat earth, on sand formations near lakes or rivers, or on sandy soil along seashores, and frequently in the hollows of undulating ground. It is promoted by low temperature

Fig. 74. — Grey Spot of Oats — *Scolecotrichum graminis.* (*a, b, c, d,* The Author ; *e,* from *F.* Krause.)

a, Early, and *b,* later stage of the disease ; *c,* shreds from a bundle of fungus threads; *d,* conidia ; *e,* an almost dead leaf, transverse section.

during the early period. It has often become worse after the soil has been marled or strewn with lime. It is said to be prevented by sulphurous ammoniac.

Different sorts of oat suffer in a different way. The black Tartarian Oat (*Avena orientalis*) has proved to be the most susceptible, and next to it come "Improved Dala Oat" (Swedish) and the yellow-corned Plume Oat, "Jaune géant à grappes." To those that have the greatest resistance belong Mesdag, Duppauer, German Moss Oat, and others.

The disease has been known both in the south and central parts of Sweden for a considerable time, but no special attention has been given to it. It is also reported from Denmark and Germany, especially in the years 1908 and 1909 from Posen and West Prussia.

Against this frequently disastrous disease no other remedy at present is known than to abstain from cultivating oats on a diseased field.

Somewhat similar spotted diseases have been noticed on Timothy-grass, **Scolecotrichum graminis f. Phlei**, on *Milium effusum*, *S. gr. f. Milii*, and on *Avena elatior*.

To this group of fungi belong the following :

Sphærella exitialis and **S. basicola**, which appear on the leaves and sheaths of wheat, rye, and barley. They produce long, pale stripes with small brown dots.

S. recutita causes long, parallel, grey stripes, on which are black dots arranged in rows, on the leaves of cock's-foot-grass (*Dactylis glomerata*).

S. isariphora produces pale, later on black, spots on leaves and stem of spurry (*Spergula*).

Læstadia microspora has been seen on oat in Denmark. The sheaths, especially the lower ones, were covered with black prickles. The plants become small, and the panicles develop poorly.

Sphærulina Trifolii yields on the upper side of the leaves of white clover (*Trifolium repens*) plenty of small, globular, pale brown spots, surrounded by a purple-reddish brim, and covered with small black dots. This fungus has been observed in Denmark and Germany.

Pleosphærulina Briosiana causes spots on the leaves of lucerne (*Medicago sativa*), some of which are small and brownish-red, others larger and leather-brown ; the latter being mostly at the points and edges of the leaves, and containing the spore-cases of the fungus. This fungus is reported from Austria.

CHAPTER XI

NECTRIACEÆ

THE wall of the spore-case is of a soft consistency, flesh-like or filmy, and highly coloured red or yellow. The spore-cases are globular or flask-shaped, partly submerged in an equally soft base, and have a papillary opening.

Potato Winter-Rot. (*Nectria Solani; Fusarium Solani.*)

This disease is very common on potatoes that are stored up for the winter. It appears in the shape of either convex or concave small blotches on the skin, the blotches being covered with white or pale red fungus tassels. These tassels consist of mycelium threads, which bear conidia. These are of two sorts, either globular (*Monosporium*), or spool-formed, cross-jointed and slightly curved (*Fusarium*). At the germination of the latter there appears a generation of spores enclosed in an envelope (*Cephalosporium*). The entire inside of the severely attacked tubers is soon transformed into a soft fetid mass. But on tubers which are only slightly touched by the disease, and which might during the following year be used as potato seed, a second stage of development of the fungus will evolve in the form of small, aggregated, pale ochre coloured, or sometimes crimson, papillose spore-cases on the skin. The papillæ contain numerous spore-bags, each one with eight two-celled spores, which, freed from their cover, will germinate and then infect the fresh young potato-tubers. As a rule, the disease is not discovered at the potato harvest; it is then still latent and develops later on.

This fungus was first described in the year 1879, and was then named *Nectria Solani*. It was then considered to be one of those fungi which now and then might appear on tubers that are decaying, but

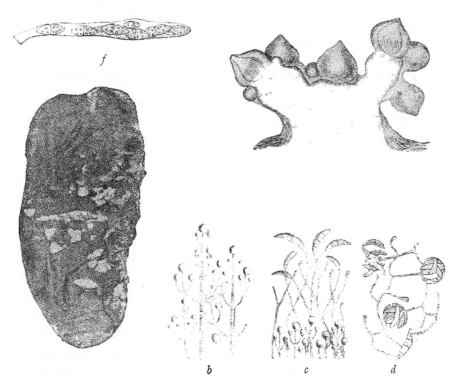

FIG. 75.—POTATO WINTER-ROT—*Nectria Solani.* (*a*, FROM BOARD OF AGRI-CULTURE (LONDON); *b*, *c*, *d*, FROM G. MASSEE; *e*, *f*, FROM J. REINKE AND G. BERTHOLD.)

a, Potato with fungus-bearing blotches; *b*, *c*, *d*, different forms of conidia : *b*, *Monosporium ; c, Fusarium* (the upper part of the figure); *d*, *Cephalosporium ; e*, spore-case ; *f*, spore-bag.

that are powerless to attack, or kill, perfectly sound and whole tubers. It is only recently that this fungus has changed its character.

The disease was malignant in Germany and France in the year 1907. A great deal of the potato seed, which in the year 1908 was exported from those countries to the Transvaal, was so badly affected with this disease that the Government of the Transvaal in the same autumn issued an ordinance that henceforth all potatoes that were attacked by the disease to the extent of 1 per cent. should be either destroyed or returned to the exporter.

PROTECTIVE MEASURES.—(1) Take care that the potato crop is dry before storing up for the winter, and that the storehouse is sufficiently dry and well ventilated. (2) Examine the condition of the stored-up potatoes now and then during the winter ; should diseased tubers be found, these must instantly be removed and destroyed. (3) Do not use for fodder slightly diseased potato, without previous thorough boiling. (4) Do not use as seed any potato that is affected by the disease, be the attack ever so slight. (5) Do not plant potatoes in a field whence, during the previous three or four years, a diseased crop has been taken. (6) If there has been disease the soil should be strongly fertilized with lime.

Snow Mould. (*Nectria graminicola; Fusarium nivale.*)

In the spring, when the snow melts, and especially after very snowy winters, and if the snow has settled without previous black-frost, there will frequently be found in the months of March or April a web-like mesh, either white or reddish-grey, spread over the sprouts and leaves on the fields of the winter crop and also over pastures. The sprouts lie prone, and the plants will shortly rot away altogether unless a drying-up wind soon sets in. Empty patches of different sizes form over the fields. The slower the snow melts, the more the mould will extend, and the greater the destruction. This mould consists of a tangled maze of jointed threads. In some places these threads develop conidia, which in their turn germinate and extend this mesh further. In this stage of development the fungus is named *Fusarium nivale*.

The disease can be brought to the plants through the soil, if this contains the mycelium of the fungus, but mostly it comes from the sowing-seed, where it has been detected in the shape of a latent mycelium, spread underneath the cuticle.

About a week or so after the mouldy leaves have withered away, there will appear on them a continuation stage of the fungus, taking the form of small groups of little, black, round spore-cases, arranged along the nerves. Each one of them has a circular opening at the top. This stage of the fungus is called *Nectria graminicola*. Towards the blooming time of the seed, the tiny corn substance is infected, when the spore-cases ripen, and they let out the spores.

The disease troubles rye especially, and wheat next. Different varieties of the same sort of seed suffer in a different way. Zealand-

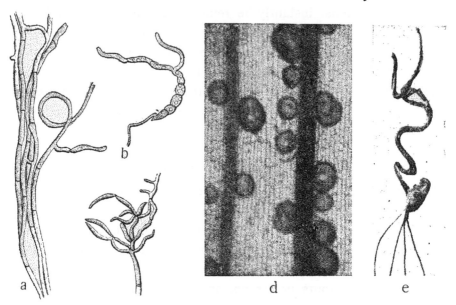

Fig. 76.—Snow Moult—*Nectria graminicola*. (*a*, *b* AND *c*, FROM P. SORAUER ; *d* AND *e*, L. HILTNER AND G. IHSSEN.)

a, *c*, *Fusarium nivale*; *a*, mycelium threads with a sling ; *b*, conidia with germination-tubes ; *c*, the origin of the conidia ; *d*, *Nectria graminicola* on a dead rye-leaf ; *e*, a germinated diseased rye-corn, unable to reach the surface.

rye has proved to be one of the most susceptible, while Petkus-rye has shown great resistance.

PROTECTIVE MEASURES.—(1) Assist the speedy melting of the snow by distributing the snow-drifts over the fields and pastures. (2) Drain off quickly the water that accumulates through the melting of the snow. (3) Harrow the earth lightly in the spring as early as possible. (4) If the sowing-corn is under suspicion of carrying infection, it should be steeped in a sublimate solution (1 in 1,000). This solution should be prepared in wooden vessels, not metal, and the proportions should be calculated as 10 litres or 2 gallons solution for 1 cwt. of corn. The corn is spread in a heap on the floor of a barn and the solu-tion is slowly poured over it, while it is stirred up all the time with a wooden shovel. Every corn must get its share of the moisture all over. The procedure takes at least some-thing like half an hour, and for larger quanti-ties a longer time; for instance, one hour for 20 cwt. Not later than one hour and a half after steeping, the corn is spread out thinly to dry. Whilst drying, it should be thoroughly stirred. It might be sown the next day, if sufficiently dry to pass through the sowing-machine. As the sublimate is ex-tremely poisonous, the

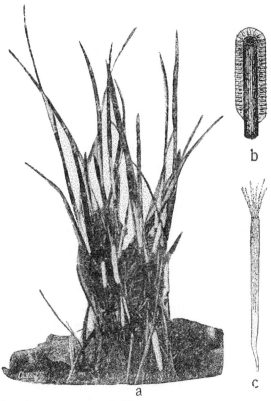

Fig. 77.—Reed Mace Fungus of Timothy Grass —*Epichloë typhina*. (*a*, From K. von Tubeuf; *b* and *c*, from E. Rostrup.)

a, An entire tuft, most of the straws diseased; *b*, longitudinal section of a spadix, with numerous flask-like spore-cases along the sides; *c*, spore-bag with thread-like spores.

barn floor, as well as all vessels and implements that have been used, should be carefully flushed with water.

Reed Mace Fungus of Grass. (*Epichloë typhina*.)

This disease appears on grasses, cultivated as well as wild, as, for instance, *Phleum pratense*, *Dactylis glomerata*, *Poa trivialis*, and *P. nemoralis*, *Agrostis*, *Holcus*, and *Anthoxanthum odoratum*. It looks like a cover over the topmost sheath of the straw. This cover is first greyish-white and later on golden-brown ; somewhat similar in hue to the female ear of cat-tail (*Typha*), from which the species name of the fungus is derived. The disease stops the growth of the straw and prevents the formation of ears. The greyish-white cover consists of closely crowded fungus threads, extending radially and producing conidia at their points. These conidia are unicellular.

As the cover acquires a golden - brownish tint, these fungus threads are crowded out by other fungus threads, which when fully developed turn out to be spore-bags, each one containing eight long and narrow jointed spores. It is surmised that the conidia diffuse the disease during the summer from one plant to another, while the actual spores prolong the life of the fungus from one year to another.

Occasionally the disease has been very malignant on Timothy grass. This was the case in Sweden in the years 1901, and especially 1902, when it appeared to a large extent over many of the central and to-wards the southern provinces. In several places the harvest was reduced to only a third part of an average output. During the same time, however, the disease was insignificant on cock's-foot grass and other grasses in the same localities.

It seems as if this fungus can retain its vitality in the rhizomes of the perennial grasses, and might possibly also be diffused by means of the seed. No remedy is known.

Ergot Disease of Grasses. (*Claviceps purpurea.*)

From bygone days certain sorts of excrescences, that occasionally appear in the ears and panicles of cereals and grasses in place of normal corn, have been known by the name of Ergot. They are several times as large as ordinary corn, and there may be many on the same ear or panicle. They are hard, pure white inside and dark violet coloured outside. They lodge preferably on rye, and were named *Secale cornutum* by the early botanists.

Ergots are the sclerotia intended for the hibernation of the fungus

Claviceps purpurea. Towards harvest-time they easily get loose from the ears and fall to the ground. The remaining sclerotia go with the corn into the barn and their stuff gets mixed in the threshed corn.

If the ergots are lying in the open during the winter and covered by the earth to a depth of about 1 centimetre, about ⅓ inch, they will germinate in the spring. From each ergot there will then grow out several fruit-bodies, frequently ten or more. These consist of a thin reddish-yellow, fragile stem and a globular, darker red head. The stem becomes long enough to bring the head above the surface of the ground. All round the outside of the head are embedded flask-shaped spore-cases, containing long spore-bags, each one having eight long, acuminated spores. These spores

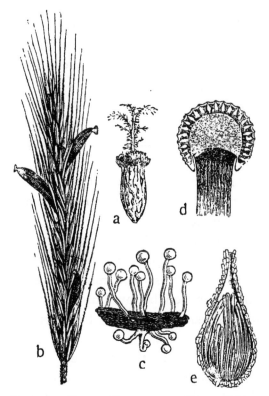

FIG. 78.—ERGOT DISEASE OF RYE—*Claviceps purpurea.* (FROM E. ROSTRUP.)

a, Rudiment of ergot, germinating conidia on its surface; *b,* rye-ear with three ergots; *c,* ergot with thirteen out-grown fruit-bodies; *d,* head of a fruit-body, intersected, flask-shaped imbedded spore-cases all round the surface; *e,* spore-case with spore-bags.

become released at the blooming time of the cereal, or the grass. The wind carries them to the surrounding fields, and some settle in the open blooms, stick to their stigmas, germinate, and grow down to the corn embryo. A single ergot is sufficient for the contamination of thousands of rye-blooms.

The infected corn embryos develop in a different way to that of the sound ones, and exude from their surface a sticky, sweetish, pale yellow fluid, somewhat similar to the honey-dew exuded or secreted by the plant-louse; it is in consequence frequently called "honey-dew." This fluid contains numberless small bud-cells, which are the form of the summer spores of the fungus. For some time this stage of development was considered as a special sort of fungus, and called *Sphacelia segetum*. By means of the bud-cells the fungus is readily diffused to the sound blooms in the same ear or in other ears. The spread takes place through the exuded fluid, by the wind shaking the ears towards each other, or through insects visiting bloom after bloom.

After a few days, however, the exudation of bud-cells will cease. The infected corn embryos grow out and finally assume the appearance of ergots.

Among the cereals it is mostly rye that is troubled by this pest. Occasionally 10 per cent. of the rye-ears are beset by the ergots. Rye sown early is more immune than that sown late. This fungus is also found on barley, but not so often as on rye, and it rarely attacks wheat. On oats it has been seen in Algeria and in North America.

In recent times the discovery has been made that within this kind of fungus there exist several specialised forms biologically different. Those that have been discovered up to the present time are the following: (1) A form on rye and barley and on the following grasses: *Hordeum murinum, Dactylis glomerata, Avena elatior, Festuca elatior, Calamagrostis arundinacea, Baldinguera arundinacea, Briza media, Anthoxanthum odoratum*, and others; (2) a form on *Glyceria fluitans*; (3) a form on *Poa annua*; (4) a form on the English rye-grass (*Lolium perenne*); and (5) a form on *Milium effusum* and *Brachypodium silvaticum*.

The last one is peculiar in that it develops numerous bud-cells, but only rudimentry sclerotia, on the early blooming *Milium effusum*; whilst on *Brachypodium silvaticum*, which blooms later in the summer, and to which the fungus is carried from *Milium effusum*, it not only develops bud-cells, but also fully developed sclerotia.

Ergots are poisonous. Rye bread much infested with ergots produces a disease called *Raphania*, that sometimes can be very malignant. The symptoms are syncope, convulsions, and lameness. Epidemics have been known when only 5 per cent. of the people attacked have been saved from death. Cattle and poultry have also suffered after consuming ergot - mixed corn. But when the corn has been stored up for some time the effect of the poison lessens and finally ceases. As a safeguard the ergots should be separated from the corn before it is given as fodder or ground to flour.

PROTECTIVE MEASURES.—(1) Pick the ergots from the ears while the crop still stands, and gather them in a bag. They are rendered harmless by burying them at least 20 inches or ½ metre deep, and keeping them covered up for a whole year. Or they might be offered to druggists for the preparation of medicines. (2) Separate the ergots or portions thereof by means of a sieve from the corn of cereals and grasses to prevent next year's crop from being infected, in case the corn be used for sowing, or otherwise to protect both people and animals from its poisonous effects, should it be used for food. (3) If the corn of cereals or

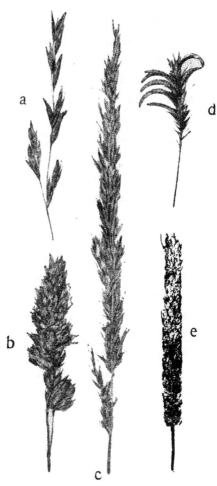

FIG. 79.—ERGOTS OF GRASSES.
(THE AUTHOR.)

a, Festuca arundinacea ; b, Dactylis glomerata ; c, Calamagrostis arundinacea; d, Triticum desertorum ; e, Phleum pratense.

grasses is so mixed with fragments of ergots that they cannot be separated, then the whole should be poured into a vessel containing a 32 per cent. solution of chloride of potassium, and stirred for a quarter of an hour, while the fragments are skimmed off with a scoop, as they are lighter than the corn and float. Then the corn is rinsed at once with fresh water and spread out to dry. (4) Plough the stubble in the autumn so deep that ergots, which before or at the harvest have fallen to the ground, are prevented from germinating, and take care that they are not brought nearer the surface in the spring ploughing, as they have not been destroyed by being ploughed down, although their development has been arrested. (5) As the ears are only susceptible to infection during the time they are in bloom, it is advisable to try to induce an even and short blooming period of the cereals. This is done by spreading the manure carefully and evenly, by sowing the corn as far as possible on the same level, and by being careful not to sow both early and late blooming sorts of the same cereal close to each other. (6) If there should be reason to suspect that the grasses growing alongside ditches carry ergots, then these grasses should be cut during their blooming period.

Another form of the same genus is **Claviceps microcephala** on *Phragmites communis*. This form might also be transplanted on *Molinia cærulea*, *Aira cæspitosa*, and *Nardus stricta*.

CHAPTER XII

DOTHIDEACEÆ

THE spore-cases are imbedded in a black, crust-like foundation, and have no regular walls.

The damage done by fungi of this group attacking agricultural plants is generally of minor importance. **Phyllachora graminis** forms long, black, shiny spots on the leaves of *Dactylis glomerata,*

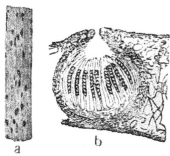

FIG. 80.—PHYLLACHORA GRAM-
INIS. (*a*, THE AUTHOR; *b*,
FROM A. B. FRANK.)

a, A piece of a leaf of *Triticum caninum* with fungus spots; *b*, section of imbedded spore-case, with numerous spore-bags.

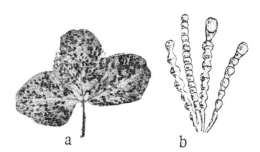

FIG. 81.—PHYLLACHORA TRIFOLII. (*a*, FROM E. M. FREEMAN ; *b*, FROM E. ROSTRUP.)

a, Leaf with fungus spots ; *b*, bead-like, jointed, filamentous tubes.

Agrostis stolonifera, and other grasses. Eventually the spots will show on both sides of the leaves, which turn yellow and soon wither.

In this same genus are generally included two other parasitical fungus forms, on account of the similarity in their appearance, although no actual spore-cases have been detected—only bud-cells in moniliform rows. These two forms are the following :

P. Trifolii, on the leaves of red, white, and alsike clover, where it forms small, crust-like spots, black on the under side and brown on the upper, with pear-shaped, bilocular, brown conidia (*Polythrincium Trifolii*); and **P. Pastinacæ,** which generates similar crusty patches on the leaves of parsnip.

Diachora Onobrychidis forms black, crusty patches on the leaves of esparcet and on *Lathyrus tuberosus*. The spots develop partly pycnidia, partly genuine spore-cases.

CHAPTER XIII

PEZIZACEÆ

THE fruit-body of the fungus, which in these fungi is called *apothecium*, is soft, fleshy, or waxy, calyx- or cup-like, usually extended from the substratum by a shorter or longer stipes. On the upper side of the concave spore-cup are numerous spore-bags, placed side by side and mingled with narrow succulent threads.

Sclerotinia.

Fungi of this form produce hard, irregular, black sclerotia, which arise either on the exterior or interior of the plants attacked. The sclerotia vary from the size of a cabbage-seed to that of a hazel-nut. From them grow out the soft fruit-cups after resting periods of different lengths.

Stem Mould.

(*Sclerotinia Fuckeliana ; Botrytis cinerea.*)

This fungus affects a great number of agricultural as well as garden plants, the chief ones attacked being potatoes, rape, turnip, carrot, peas, vetch, and lupin. It looks like an ashy-grey mouldy nap on the lower and middle parts of the stem and also on the leaves. It seldom appears higher up on the plant, although sometimes on pea-pods. The affected parts are stunted in their development, and severely attacked plants hardly produce blooms and fruit. From the mould-cover arise numerous threads, much ramified at the summit, where they set free a large number of small, closely clustered, ovate conidia. This stage of development is often called *Botrytis cinerea*. By means of the conidia the

133

fungus is spread during the vegetative period, and affects plants of different kinds.

The sclerotia break out mostly on the surface of the dead parts of affected plants, and form there long crusts, often thin and flat. The fruit-cups, one or a few from each one of the sclerotia, are on stipes about 1 centimetre or ⅓ inch long, and have a disc in shape like a watch-glass.

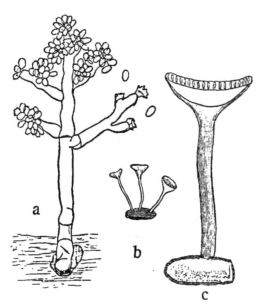

It is possible that this species of fungus embraces several biologically different forms, adapted for different sorts of host plants. Several observations prove it to be likely that the disease can also spread from place to place by means of the seed.

From Ireland comes the report of a potato-disease apparently belonging to this group—

FIG. 82.—STEM MOULD OF RAPE — *Sclerotinia Fuckeliana.* (*a*, FROM A. B. FRANK ; *b* AND *c*, FROM E. ROSTRUP.)

a, Conidia producing mycelium ramification ; *b*, sclerotium with three fruit-cups ; *c*, section of sclerotium and fruit-cup.

Botrytis on Potato. During the summer there appear on the green potato-stalk, inside as well as outside, small, black, elongated, slightly raised, crusty stripes. These crusts consist of a close mesh of fungus threads, and develop into sclerotia. When these germinate, small tufts of spore-producing fungus ramifications grow from them and are classified as belonging to the genus *Botrytis*.

PROTECTIVE MEASURES. — (1) Separate and destroy all the sclerotia. (2) Do not use sowing-seed taken from a diseased crop. (3) Avoid swampy localities, and do not place the plants too close to one another.

Root-Crop Rot. *(Sclerotinia Libertiana.)*

This disease appears on roots that have been stored up for the
winter, either in cellars or in large heaps covered with earth.   It

a   b

FIG. 83.—ROOT-CROP ROT OF FODDER BEET—*Sclerotinia Libertiana.*
(THE AUTHOR.)

a, Beet with numerous sclerotia on the surface of the lower part of the root;
b, section of beet, showing the extension of the rot inside the root.

troubles beet, rape, turnip, carrot, celery, parsley, chicory, and other
plants.   So long as the plants grow, there is little or no sign of the
disease.   There might, however, be sporadic cases, when it attacks

the growing plant—for instance, on potato, rape, carrot, and parsley—and these cases require special attention, as such plants may readily convey the disease to the storage ground.

FIG. 84.—ROOT-CROP ROT OF CARROT — *Sclerotinia Libertiana.* (THE AUTHOR.) Numerous sclerotia inside the root.

An especially interesting case with potatoes is reported from Ireland. The disease, which is called " Stalk-Disease," and also " Sclerotium-Disease," " White Spot," " Falling at the Butt," or " Haugh-ing," appears early in July on the outside of the potato-stalk, mostly on the lower part, in the shape of small, snowy-white, nappy papillæ, from which silvery drops of a watery fluid soon ooze out. The papillæ become hard, and develop into black sclerotia of a waxy nature. Similar sclerotia appear inside the hollow part of the stalk. A week after the first sign of the disease the stalk of the specimens first attacked becomes fragile and breaks off. Those sclerotia that are outside loosen easily and drop to the ground; as also do the sclerotia which are located in the hollow of the stalk, when this withers down in the autumn. The sclerotia rest in the earth during the winter, and germinate the following summer in the shape of calyx-like spore-cups with long stipes; and then the wind carries the infecting spores to the green potato-stalks.

On several plants the disease renders itself conspicuous by a white, nappy, luxuriant mycelium, extending more and more over the surface of the root, and possibly also on the stalks and leaves, if any remain. The growth and extension of the mycelium is encouraged by a damp atmosphere; hence a greater destruction takes place in cellars than in the earth-heap.

If straw be placed in an earth-heap about the roots for a covering, this will rather increase the ravages of the scourge than otherwise. By means of the mycelium the disease diffuses rapidly from root to root. Inside the root the rot is also extending and the fibres decay. The mycelium may live year after year in the floors of cellars or in cracks on walls and ceilings, etc. Hence the disease repeats itself in the cellars where it originally appeared, even although fresh, esculent roots are stored up there.

In the greyish covering at an early date appear black, hard sclerotia of differing size and form; the size varies from that of a pea to that of a bean. Many of them are clustered together. The form follows mostly that of the substratum. On the thick roots, and also

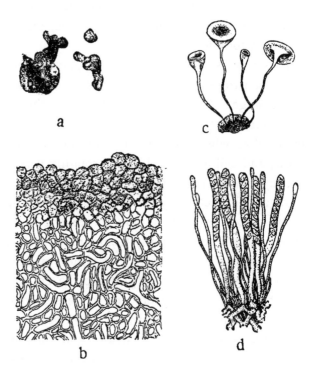

Fig. 85.—Root-Crop Rot. (*a* and *d*, From R. E. Smith; *b* and *c*, from E. Prillieux.)

a, Sclerotia; *b*, exterior border of an intersected sclerotium; *c*, sclerotium with four growing spore-cups; *d*, spore-bags with sterile threads.

in hollows of stalks and roots, they become round and thick, with numerous irregular sinuosities. On leaf-stalks and leaf-nerves they are flat and ovate-oblong. By means of the sclerotia the fungus can survive for years in a state of repose, should untoward conditions prevent its normal growth.

In the spring the sclerotia develop fruit-cups, usually several from each sclerotium. They are borne on a tubular stipes up to 3 centimetres, or about 1 inch in length. The disc is concave, the centre of the concavity being connected with the tube of the stipes. In this way the fruit-cup has the appearance of a trumpet. The upper surface of the disc is formed of crowded spore-bags, each of which contains eight egg-shaped spores. When the spores germinate, a mycelium develops; this, on a suitable basis, grows into a cottony felt.

PROTECTIVE MEASURES.—(1) Separate carefully all diseased roots at harvest-time, even those only suspected, and do not let them get into storage. (2) Gather and burn all the sclerotia that might be ready formed at the harvest-time. (3) If the crop is to be stored in a cellar where the disease has previously been, then the floor, walls, and ceiling should be thoroughly cleansed, lest some remaining mycelium infect the roots; and to be quite sure, it is better to fumigate the cellar by burning powdered sulphur in an iron vessel, to be placed on a flat stone, to prevent danger from fire. (4) Examine the stored-up supply from time to time, removing everything that is diseased, and destroy it either by burning or by burying it deep below the surface. (5) Keep the cellar ventilated as much as possible. (6) If the roots are to be stored in an earth-heap, then they should first be covered by a thin layer of fresh soil, upon which is placed a layer of dry straw, and then earth again upon that. (7) Do not use roots for fodder, even if only slightly diseased, without previous boiling. (8) In a field that has yielded a diseased crop, do not raise any sort of plant that is susceptible to the disease until three years have elapsed.

Clover-Sclerote. *(Sclerotinia Trifoliorum.)*

This disease becomes conspicuous in the autumn on young clover that sprouts after the spring sowing. On the leaves of sundry plants appear bleached brown spots that spread rapidly. Soon both leaf and stalk wither. But at this period the disease usually escapes the planter's attention. However, at the melting of the snow the following spring it is clearly evident: a larger or smaller amount of clover-plants lie prone and withered. On the field there appear large, bare patches, and in severe cases the entire clover-plants die off. On these dead plants can be seen black, irregularly formed, and loosely attached sclerotia, usually round the neck of the root.

The fungus causes great devastation on red clover, but has also been seen on alsike clover, white clover, crimson clover, and on esparcet, lucerne, and *Medicago lupulina*, and several other plants of the same family. With the red clover the sclerotia are the size of a pea, or somewhat larger, and develop on the root. With the alsike clover they are smaller, and appear the entire length of the stalks above ground. With the *Medicago sativa* the fungus causes the upper part of the root to rot, and the sclerotia develop round the neck of the root. But with regard to this it frequently happens

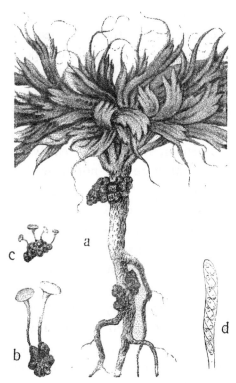

FIG. 86.— CLOVER-SCLEROTE—*Sclerotinia Trifoliorum.* (THE AUTHOR.)

a, Clover-plant killed by the fungus, sclerotia on the root; *b* and *c*, sclerotia with grown fruit-cups; *d*, spore-bag.

that plants which have more stamina can revive on account of the long, deeply-set roots. Beneath the decayed spot buds arise, which during the summer may develop sprouts, and appear above the surface.

Late in the autumn there grow out from the sclerotia pale red fungus cups, borne on long, narrow stipes, which frequently in their formation are irregularly spiral. The length of the stipes accommodates itself to the depth at which the germinating sclerotium is placed. The stipes becomes exactly the length required for the purpose of bringing the disc above the surface. On the upper side of this disc are closely crowded spore-bags, each of which contains eight spores. The spores germinate at once, and originate a mycelium. This attacks the leaves, stalks, and roots of the clover-plants. It forms a mesh that penetrates the whole plant, and kills it altogether by the spring, perhaps earlier.

A mild winter, abundant moisture, and crowded plants promote the development of the disease. Generally the attack is most serious the first year after sowing, presumably because the plants are then more tender, hence more easily accessible to an attack from the fungus. Manure has not proved to have any checking effect upon the disease, but rather the contrary. Even a strong dressing of lime has proved futile.

PROTECTIVE MEASURES.—(1) If the plants that die in the spring are only few and scattered, then they should be dug up with all the sclerotia upon and around them, and fresh seed sown in the bare patches. (2) Should most of the clover-plants in a field, consisting of nothing but clover, die off, it will be best to plough up the ground and sow some mixed seed that will yield green fodder, thus finding some return for the lost clover crop. (3) For a period of three to four years no clover should be planted on a diseased field. (4) If there be the slightest fear that the disease may appear, then clover should not be sown alone, but mixed with grass-seed. (5) As it seems quite likely that the disease may be spread by means of the sowing-seed, it is advisable to obtain the seed from a field that has proved to be free from the disease.

Closely related to this is **Sclerotinia Nicotianæ,** which brings whitish spots, and later on black sclerotia, upon leaves and stalks of tobacco-plants.

To this group belongs also **Pseudopeziza Trifolii** ("Clover Leaf Spot"), which infests clover, lucerne, *Anthyllis Vulneraria,* and *Lotus.* The fungus produces on the upper side of the leaves a plentiful supply of small, dark, round spots. Upon these spots arise, later on, waxy discs with a rugged edge. The disc is covered with crowded spore-bags, intermingled with narrow succulent

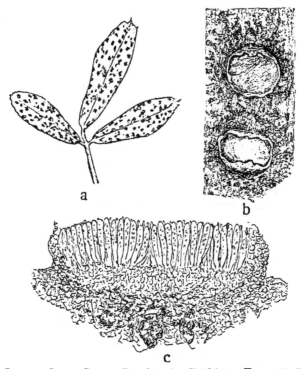

a

b

c

FIG. 87.—CLOVER LEAF SPOT—*Pseudopeziza Trifolii.* (FROM E. PRILLIEUX.)

a, Leaf with fungus spots; *b,* two fungus cups with outward turned or reflexed edge; *c,* a fungus-cup section, with numerous closely crowded spore-bags.

threads. In severe cases all the leaves may prematurely wither and die. The disease is somewhat checked if the parts most diseased be cut off.

HELVELLACEÆ

THE fruit-cups are fleshy, cap- or club-shaped.. The outside of the fungus head is covered all round by a layer of crowded spore-bags, mixed with sterile threads.

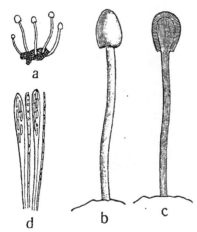

FIG. 88.—MITRULA SCLEROTIORUM. (FROM E. ROSTRUP.)

a, A sclerotium with five outgrown fungus cups ; *b*, a fungus cup ; *c*, a similar one, in section ; *d*, spore-bags from the cover at the top of the fungus cup.

Among these fungi is **Mitrula Sclerotiorum,** which occasionally appears on clover, *Lotus*, and other related plants. The effects of the attack are like those of the clover-sclerote, and, like that fungus, it develops sclerotia. From the sclerotium grow, after a certain resting period, five or six, occasionally as many as thirty, fungus cups. These are at first white, then of a fleshy tint ; their stalks are narrow and the head swollen, like a club.

This fungus is as common in Denmark as the stem-rot on clover.

CHAPTER XV

SPHÆROPSIDEÆ

THESE fungi have their conidia inside special globular pycnidia, visible to the naked eye only as small brown or black spots. These pycnidia are often provided with a small round opening, through which the conidia obtain their exit in a mass of serpent-like slimy threads.

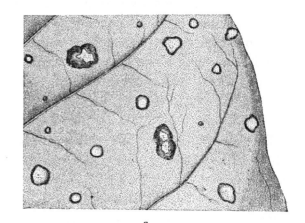

a

Phyllosticta.

The pycnidia form well defined, dead spots on the green leaves. These spots are often surrounded by a dark red or brown brim, and the pycnidia appear as brownish - black specks at the middle part of the spot. The conidia are egg-shaped, colourless, and unicellular.

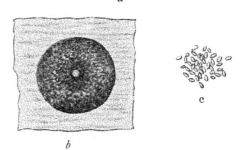

b

c

FIG. 89.—PHYLLOSTICTA TABACI. (FROM O. KIRCHNER AND H. BOLTSHAUSER.)

a, Portion of a tobacco-leaf with fungus spots ; b, pycnidium from the centre part of a spot c, bud-cells.

To this belong a great number of species practically different as regards their parent plants. **Phyllosticta Betæ** forms on the leaves of beet round, yellow patches, pale at the centre, with a darker brim, occasionally with dark spots at the centre.

Similar blotches on the leaves of tobacco are caused by **P. Tabaci,** and on clover by **P. Trifolii,** on lucerne by **P. Medicaginis,** on Windsor-bean by **P. Fabæ,** on rape and turnip by **P. Brassicæ,** on buck wheat by **P. Polygonorum,** and so forth.

Phoma.

The pycnidia appear on stalks and roots—not on the leaves. The diseased blotches are not limited. The conidia are mostly unilocular.

Carrot Disease. (*Phoma Rostrupii.*)

This fungus puts in an appearance towards the end of the summer at the upper part of the root, either round the leaf-rosette or on the top of the root just below the surface. It appears as concave grey blotches, sprinkled with small, blackish papillæ. These papillæ are pycnidia, containing a great number of egg-shaped conidia, kept together by a slimy fluid. When the weather is wet the ripe conidia pass out through the opening of the pycnidium as a long, flesh-coloured thread.

Diseased roots, stored up for the winter either in cellars or earthen heaps, get worse, and diffuse the disease to those roots that were sound when stored.

This fungus will cause great damage to the seed business of the carrot-planters. No seed can be obtained from a carrot, be it ever so little touched by the disease. The fungus travels from the root upwards in brown stripes on the stalk. These stripes are similar to those which appeared during the previous year on the root, and they are covered with pycnidia. When the plant is ready to bloom, the root is already rotten. The stalk withers, and the plant dies without yielding any seed.

For many years the disease has spread much in Denmark and caused great losses. Different sorts of carrots are differently attacked. The devastations have proved worst on slightly sand-mixed soil.

PROTECTIVE MEASURES. — (1) Remove and destroy at harvest-time all diseased roots. (2) Inspect the storage frequently, and

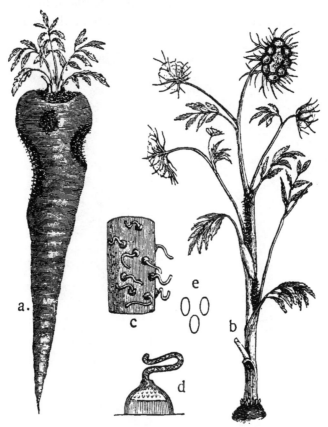

FIG. 90.—CARROT DISEASE—*Phoma Rostrupii.* (FROM E. ROSTRUP.)

a, Carrot with concave blotches, covered with pycnidia ; b, the same on a plant in bloom ; , portion of a stalk with conidia strings ; d, section of pycnidium ; e, bud-cells

remove anything that is diseased. (3) Use perfectly sound roots for seed-production (4) Select for cultivation those which have shown the greatest resistance. (5) Use seed only from a sound field. (6) Cultivate the roots in nutritious and well-prepared soil. (7) Do not plant carrots in infected fields for several years.

Cabbage Canker. (*Phoma oleracea.*)

This disease appears under somewhat various forms in different seasons.

One form appears on the growing cabbage-plant, and causes the death of the trunk-root. All the softer parts of the root dissolve ;

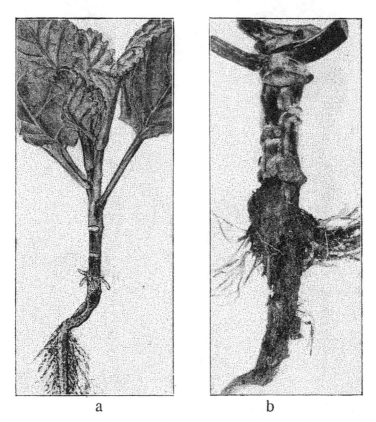

a   b

FIG. 91.—CABBAGE CANKER—*Phoma oleracea.* (FROM J. RITZEMA BOS.)

a, Young cabbage-plant at the early stage of disease ; *b*, older plant, with scars from fallen leaves and numerous freshly-grown branch roots.

only the firmer fibres remain. From the trunk, beyond the dead part, there usually grow out plenty of branch roots. If these grow strongly, then the cabbage-plant may survive, at any rate for a time ; otherwise it quickly withers away. The disease is found also on very young plants having merely three or four leaves. Such plants are conspicuous from having stiffer, straighter leaves, not curved as in the sound plant. The disease is worst on red cabbage, next on Danish trunk-cabbage, and only slightly on cauliflower. Plants attacked cannot stand erect—sooner or later they droop; hence the name of "epilepsy" sometimes given to the disease.

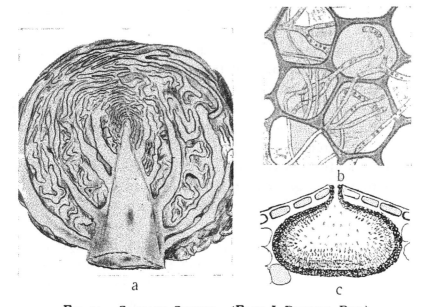

FIG. 92.—CABBAGE CANKER. (FROM J. RITZEMA BOS.)

a, Stored-up cabbage-top. with extended canker; *b*, cells from diseased tissue, with mycelium threads; *c*, pycnidium with conidia.

Another form of the disease can be diagnosed during the winter on the stored-up cabbage. Inside the cabbage-trunk are found small spots, first white, then pale brownish-grey, finally black. These canker-spots increase steadily. Usually the disease spreads through the trunk to the leaves, first the exterior, then the interior ones. Moist and warm air assists the diffusion. From the beginning of January it rapidly increases. The leaves perish and drop from the trunk. On the surface of the diseased parts is seen a fluffy mycelium, and at the same time many small pricks, first red, then brownish-black. These are the pycnidia of the fungus—*Phoma olevacea*—which are causing the disease. Inside the diseased parts the cell-tissue is interwoven with the mycelium of the fungus.

Extensive investigations have proved that these two diseases are only different forms of the development of the same disease.

But simultaneously with the ravages of the fungus, an insect— *Anthomyia Brassicæ*—is also responsible, especially with the growing plants, as it appears that the attacks of this insect open a road for the fungus.

It has not been proved that the disease follows the sowing-seed. But it has been found that seed from various cultivations produces plants that are in a different way susceptible. The disease for several years has been very devastating in the northern part of Holland (Langendijker district), where the cultivation of cabbage is conducted on a large scale.

PROTECTIVE MEASURES.—(1) Do not use plants of a sickly appearance nor those damaged by insects as seedlings. (2) Select those sorts of cabbage that are growing strongly and have proved capable of most resistance.

Closely related to this disease is **Dry Rot of Rape** (*Phoma Napobrassicæ*). It forms rot-spots round the neck of the root, and spreads down the sides of the root. The effect is that the root rots in the ground. Slightly touched roots that are stored up for the winter will further develop the disease. The rotten spots are interwoven by a mycelium that brings forth small black pycnidia. The same disease also attacks turnips.

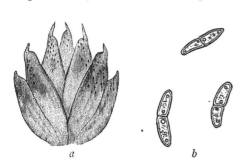

FIG. 93.—PHOMA HENNEBERGII OF WHEAT-EARS. (THE AUTHOR.)

a, Spiculæ with spots on the ear-scales ; *b*, bud-cells.

Related to this is also **Phoma Hennebergii** on wheat-ears. The ear-scales have large, irregularly formed spots, either greyish-brown or chocolate. On the spots are small dots, consisting of brownish-black pycnidia, which have a great number of jointed conidia. This disease was rather malignant in the autumn of 1899 on spring wheat near Stockholm.

Ascochyta.

The spermogonia appear on stalks and leaves. The conidia are bilocular and colourless.

To this genus belong a great number of species, differing one from another essentially by their appearance on different host plants.

Pea-Pod Spot (*Ascochyta Pisi*) causes on the leaves, stalks, and fruits, and at times on the seeds, of peas, vetch, lucerne, and others, yellow, brown-edged spots, with small brownish-black spermogonia in the centre. The disease can occasion a considerable reduction in the pea-crop.

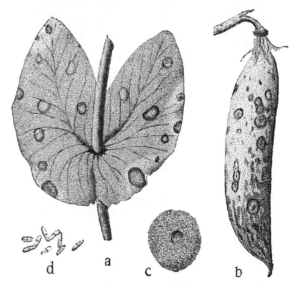

FIG. 94.—PEA-POD SPOT—*Ascochyta Pisi.* (FROM O. KIRCHNER AND H. BOLTSHAUSER.)

a, Stipulæ, and *b,* pod with fungus spots; *c,* pycnidium ; *d,* conidia.

Peas taken from diseased plants must not be used as sowing-seed.

Other species are **A. graminicola,** on cereals and grasses; **A. Fagopyri,** on buckwheat ; and **A. Nicotianæ,** on tobacco.

Septoria.

Forms belonging to this genus become conspicuous on leaves and stalks. The pycnidia contain long and narrow conidia. These are frequently provided with a row of drops or with cross-walls.

This genus also contains many species. **Grass Leaf-Spot on Cereal Sprout** (*Septoria graminum*) produces on leaves of wheat, oat, rye-grass, and *Avena elatior* pale spots with small black pricks,

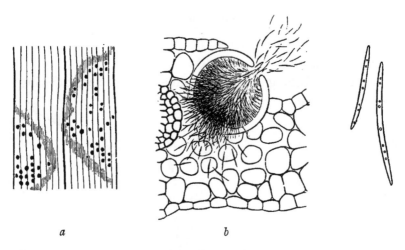

a   *b*

FIG. 95.—GRASS LEAF-SPOT OF WHEAT-SPROUT—*Septoria graminum.*
(THE AUTHOR.)

a, Portion of a leaf, with spots of the disease ; *b*, cross-section of a spot, with a pycnidium ; *c*, conidia.

arranged in rows. Occasionally it causes great destruction of the tiny wheat-sprouts in the early spring.

Similar leaf-spots are caused by **S. Pastinacæ** on parsnip, **S. Medicaginis** on lucerne, **S. silvestris** on *Lathyrus*, **S. Anthyllidis** on lady's-fingers, and **S. Spergulæ** on spurry.

CHAPTER XVI

HYPHOMYCETES

THE conidia are not enclosed in special cases, but develop either from a cup- or disc-formed bed (*stroma*), or from separated, loosely placed filamentous tubes.

(a) MELANCONIEÆ.

The conidia develop from an extended stroma, originally underneath the epidermis of the host plant, but later on laid bare, through the rupture of the epidermis.

Clover Stem-Rot. (*Gloeosporium caulivorum.*)

This disease—also called "Anthracnose"—was first noticed in America and later on in several places in Europe—viz., Bohemia, Würtemberg, Saxony, Brandenburg, and other places. It attacks red clover, and occasionally alsike clover. It forms on stem and leaf-stalks ovate-oblong, light brown spots, hollow in the centre, with a broad, black border. Frequently the leaves that are outside the diseased parts wither away.

The conidia-beds develop in these hollows. This disease has sometimes destroyed 25 to 50 per cent. of the harvest. It is worst on American clover.

A related form—**G. Trifolii**—troubles only the leaves. Other species of the same sort are **G. graminum,** that creates numerous small brown beds of conidia on the leaves of rye-grass, and **G. Dactylidis,** which causes brown papillæ on the top stalks of cock's-foot grass. Both these forms have been noticed in Denmark.

To the same fungus group belong the following forms:

Marssonia Secalis produces on the leaves of rye and barley and on several wild grasses long, greyish-white, brown-edged spots, which on their under side develop conidia-beds.

Cryptosporium leptostromiforme forms spots on the stalks of lupin (yellow and blue), at first pale, and later on brown, upon which long black conidia-beds break out. Plants that are severely attacked soon die, often before they bloom. The fungus can survive on dead lupin-stems on the ground. Hence the cultivation of lupin in the diseased soil should be avoided for two to three years.

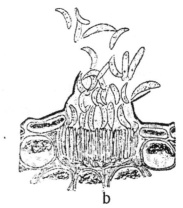

a   b

FIG. 96.—CLOVER STEM-ROT—*Gloeosporium caulivorum.* (FROM O. KIRCHNER.) *a,* Red clover plant, with long, dark spots of the disease on the stem ; *b,* section of stroma, with conidia.

(β) TUBERCULARIACEÆ.

The stroma is from the beginning placed free on the surface of the host plant, and is usually waxy or mucous.

Fusarium.

The conidia are spool-formed, slightly bent, usually provided with several traversing walls. The mucous or nappy fungus cover is first colourless, then either yellow, orange, or terra-cotta.

Mucous Mould on Cereals. (*Fusarium avenaceum.*)

During damp autumn weather there frequently appear fleshy or mucous fungoid formations, either salmon or terra-cotta coloured,

on the ears and panicles of cereals and grasses, both on awns and corns. Similar formations are found on the stubble and occasionally on the germ-sprouts of cereals. The red cover consists of a thick, felt-like layer of fungus threads, and upon this layer is a ramification of filamentous tubes, turned outwards. These bear one or more ovate-oblong spores, four to six celled, slightly curved, and pointed.

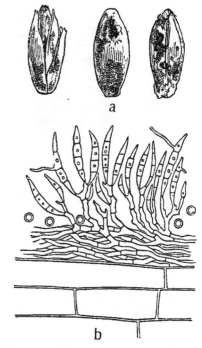

The forms which attach themselves to different cereals and grasses are often described as being each one a specific form : *Fusarium culmorum* on wheat and rye, *F. Tritici* on wheat, *F. Hordei* on barley and oat, *F. Lolii* on rye-grass, and so on.

It is asserted that seed affected by this disease is poisonous, and

FIG. 97.—MUCOUS MOULD OF BARLEY—*Fusarium avenaceum*. (FROM W. G. SMITH.)

a, Spiculæ and grain of barley with a fungoid cover; *b*, portion of the fungoid cover.

produces in man and beast the same symptoms as previously mentioned in the case of "Giddy Rye"—viz., dizziness, headache, and so forth. It has even been claimed that a special poison has been extracted which originates through the dissolution of the albumen

in the grains. However, experiments on animals with seed con·
taminated by this fungus have given contradictory results. Some
times the animals are affected, sometimes not. Hence nothing
positive can be said about it.

From rye-grains that have been strongly affected by this fungus
during the autumn there appears occasionally in the spring a
cup - like fungus, which is very similar to, and even has been
considered as synonymous with, a cup fungus named *Stromatinia
temulenta,* which has sometimes been noticed to sprout in the spring from diseased rye-grains which have been impregnated by the mycelium of another fungus, called *Endoconidium temulentum.*

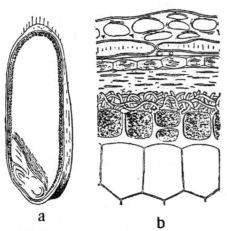

a   b

FIG. 98. — MYCELIUM IN THE SEED OF
BEARDED DARNEL—*Lolium temulentum.*
(FROM P. GUÉRIN)

a, Section of fruit ; the layer of fungus
threads (drawn in black) around the
grain, below the scale ; *b,* section of
fruit-wall.

To the cup fungus (**Stromatinia temulenta**) has also been attributed the sterile mycelium that appears in the seed of bearded darnel (*Lolium temulentum*), close against the glutinous cell - layer. This mycelium has been traced through the whole plant, but no spore-formation has been detected, nor could it be brought forth through culture. This mycelium does not cause any
disease to the host plant; but rather the contrary, as it accumulates
nitrogen for the benefit of the plant. Wherever this sort of rye-
grass appears, it is more or less accompanied by this fungoid
formation. This same sort of mycelium has been found in grass-
seed from the tombs of the Pharaohs about four thousand years
old. The mycelium renders the seeds of this grass poisonous, as
has been known since the days of Virgil, Ovid, and Pliny.

In rare cases a similar mycelium-formation has been seen in the seeds of English rye-grass (*Lolium perenne*) and also in *Lolium linicola.*

Vessel Brand on Pulses. (*Fusarium vasinfectum.*)

By the name of Vessel Brand can be called a disease which has been noticed lately on different sorts of pulses, as peas, Wind-

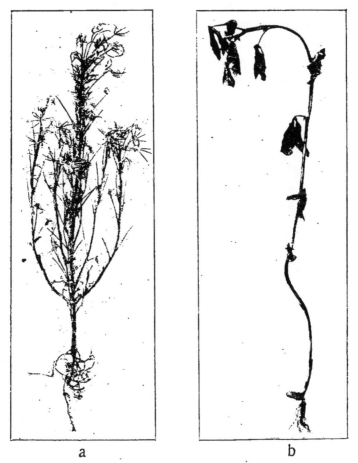

a   b

FIG. 99.—VESSEL BRAND OF PULSES—*Fusarium vasinfectum.* (FROM G. SCHIKORRA.)

a, Diseased lupin-plant ; *b*, young plant of Windsor-bean, artificially infected.

sor-bean, lupin, and others. The disease becomes conspicuous by the sudden and unexpected withering of the plants. In peas it

has been seen to start in the month of May. Sundry young leaves
and parts of the stalks become soft and pale; finally the whole
plant withers away.   This usually takes place towards the end of
June, hence this ailment has been called "the midsummer disease
of peas" ("St. Johanniskrankheit").   In the Windsor-bean a
similar destruction sets in if the plants have been infected while
very young.   But if they should be older when exposed to infection, then the parasite cannot break down the main stems, and it is
only the leaves that die.   With regard to lupin—especially *Lupinus
angustifolius*, and also *L. perennis* and *L. mutabilis*, but not *L. luteus*—
the disease has the same effect, either on young plants, shortly
before they bloom, or on older plants
which have already formed pods.
The main stems remain erect.

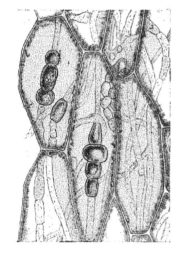

The disease begins at the neck of
the root, in one or more dark, decaying stripes, which often reach above
the surface of the soil.   The contaminating stuff gains an inlet
through sores and fissures caused
by external agency, such as the
pulling off of branch roots, and so
forth.   The fungus grows into the
tissue of the vessels, and penetrates
the whole plant.   The vessels become
filled with a yellowish mucus.   The
mycelium extends to the neighbouring cells, and develops plenty of

Fig. 100. — Cells from a Pea-
Stem, diseased by Vessel
Brand, with Mycelium and
Resting Spores. (From G.
Schikorra.)

multilocular resting spores.   These spores retain the live fungus
during the winter.   If the cell-tissue that is penetrated by the
mycelium happens to lie bare, then there appears an air-mycelium,
which develops conidia, one to two locular, or even more, say three to
six locular, by means of which the disease can be spread from plant
to plant during the vegetative period.
This disease has been noticed especially on peas in Holland

and Germany, and also on other pulses, as Windsor-bean, lupin, and clover. The form on pea has been considered as a special sort of its own—*forma Pisi.* It has, however, been found from experiments that forms from one parent plant may infect other species.

PROTECTIVE MEASURES.—(1) Do not use for sowing grains that are slow in germinating. (2) If there should be discovered places in the fields where the infection has gained a footing, then all the diseased plants ought to be taken away and destroyed. (3) Gather and destroy all diseased remains of the harvest. (4) On soil that is badly infected do not cultivate pulses for two to three years.

A similar fungoid formation—**Fusarium roseum, var. Lupini albi**—has been noticed in Germany on pods of *Lupinus angustifolius.*

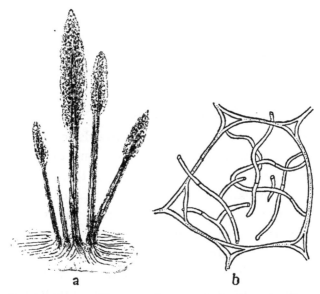

FIG. 101.—BROWN ROT OF POTATO—*Stysanus Stemonites.* (*a,* FROM J. REINKE AND G. BERTHOLD; *b,* FROM O. APPEL AND W. F. BRUCK.)

a, Group of spore-brooms; *b,* cell from the skin, with mycelium.

F. Betæ appears on beets as yellow, mucous, shrivelled excrescences on the root which otherwise appears sound. The mycelium penetrates inside the root, and occasionally forms papillæ. The disease appears on young plants at the beginning of July, when they mostly wither away. It can also cause destruction on seed-beets and on beets in winter storage.

F. Brassicæ appears in a similar way on rape, turnip, and other such plants.

Here may also be mentioned **Brown Rot of Potato** (*Stysanus Stemonites*). This parasite lives in the skin of the potato as an intercellular mycelium. From this mycelium extend greyish-black, broom-like processes that bear egg-shaped conidia at their points. The germinating tubes of these can only attack a sound potato-tuber or a potato-sprout when there is a wound by which the tubes can enter. This fungus is especially harmful, as it prepares the way for other more destructive fungi.

(γ) MUCEDINEÆ.

The conidia-bearing fungus threads are free, and protrude from the openings. The filamentous tubes and conidia are colourless.

Oospora cretacea.

This disease becomes conspicuous towards crop-time. The surface of the root is then found to be dark and to a large extent covered by a bark, crossed by fissures. Often there is an entangle-ment on the root, usually at the middle, on account of which the disease in Germany is called " Gürtelschorf "; or otherwise there are irregular cavities of various shapes.

The disease is caused by several different species of the group **Oospora** (*O. cretacea, O. rosella,* and others), often promoted by the agency of certain worms (*Enchytraeideæ*) which bore hollows in the root. The fungi are not able to attack the sound surface of the root, and can gain access only through wounds. The disease is worst after dry or open winters. It is encouraged by a wet and cold spring, or a dry and hot summer. It increases on being fertilized with Chili saltpetre.

The disease has appeared in Germany in many places since about the year 1895, being especially malignant during 1899 and 1903 in the district between Aschersleben and Hildesheim. It can reduce the crop by 25 to 50 per cent.

PROTECTIVE MEASURES.—(1) Avoid contaminating sound fields with diseased soil. (2) Drain swampy fields. (3) Sprinkle the soil with lime.

In North America a somewhat similar disease, often called " Beet Scab," attacks the sugar-beet. It is thought to be caused by

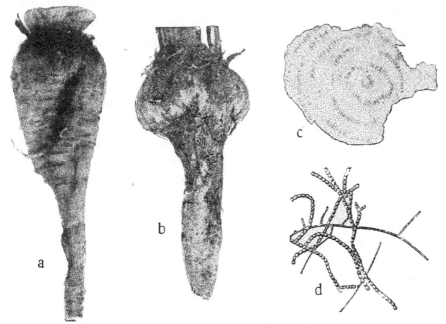

FIG. 102.—OOSPORA CRETACEA. (FROM F. KRÜGER.)

a, Early, and *b*, later stage of the disease ; *c*, transverse section of a diseased beet ; *d*, vegetative filamentous tubes and moniliform spores of the fungus.

Oospora Scabies. It begins as small excrescences on the surface of the root. These are at first either separate or in patches. Later on they extend and unite into larger or smaller knots, covered by papillæ. The disease is considered to be identical with scab on potatoes, caused by a fungus of the same name.

To the same group belong several fungi that form leaf-spots.

Ramularia Betæ produces on the leaves of beets round, whitish-grey spots, visible on both sides, and surrounded by a brown border, with white in the middle, owing to fungoid threads and conidia. This disease has occasionally been so malignant in Denmark that almost every leaf in a beet-field has been affected.

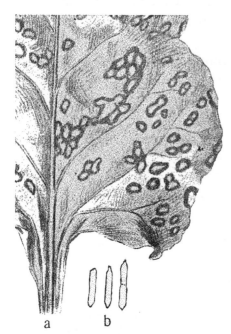

FIG. 103.—RAMULARIA BETÆ. (FROM E. ROSTRUP.)

a, Portion of a leaf with fungoid spots; *b,* conidia.

Similar spots are caused by **R. sphæroidea** on the leaves of vetch, **R. Onobrychidis** on the leaves of esparcet, and **R. Schulzeri** on the leaves of bird's-foot trefoil (*Lotus corniculatus*).

Nearly related is **Ovularia deusta,** which forms small, pale spots on the leaves of *Lathyrus.* These spots have bunches of fungoid threads.

(δ) DEMATIEÆ.

Fungoid threads that are carrying the conidia stand free. The filamentous tubes and the conidia—at any rate one of them—have coloured, light or dark brown walls; hence a sooty cover is formed on the affected parts of the plants.

Cercospora.

Fungoid threads and conidia are pale brown, the latter being pointed towards the terminals, and usually ovate-oblong or needle-like, with several cross-walls.

Cercospora concors.

From the middle of July this fungus causes irregular spots of various sizes on potato-leaves. These spots are at first yellow above, and are somewhat numerous. Beneath they are covered with a grey-violet nap of fungoid threads which cast conidia. The leaf turns more and more yellow, while at the same time the colour of the spots becomes brownish-black. The conidia on the under side are numerous, short, and blunt; on the upper

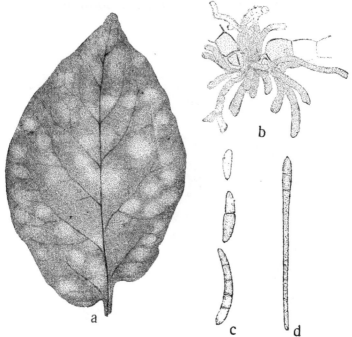

FIG. 104.—CERCOSPORA CONCORS. (FROM G. LAGERHEIM AND G. WAGNER.)

a, Diseased leaf seen from above; *b*, section of a bunch of fungoid threads; *c*, short, and *d*, long conidia.

side fewer, longer, ovate-oblong, and somewhat pointed at one end.

This disease may reduce the crop considerably, especially as it appears early in the summer. It has been noticed in Sweden in 1896 and 1902, and also in various other parts of Europe.

PROTECTIVE MEASURES.—(1) Remove and destroy all old leaves from the diseased potato-field. (2) Do not cultivate potatoes in a field that has yielded a diseased crop until two to three years have elapsed.

Cercospora beticola.

This disease begins as small, brown, red-brimmed spots, irregularly scattered over the blades of the leaves. The brown colour of the spots soon turns into grey, and then black. The spots dry up, and at the same time fissures and holes form over the blade.

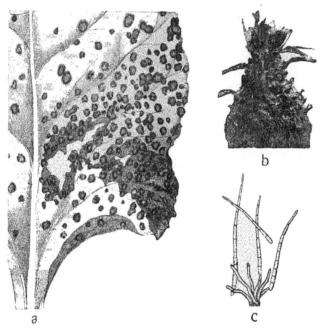

FIG. 105.—CERCOSPORA BETICOLA. (*a*, FROM O. KIRCHNER AND H. BOLTS-HAUSER ; *b* AND *c*, FROM B. M. DUGGAR.)

a, Portion of a leaf with disease spots ; *b*, the conical root-neck of a diseased beet ; *c*, a mycelium bunch, liberating conidia.

Finally, the whole leaf becomes dry and shrivelled. The outside leaves of the rosette first become diseased, then those inside. Meanwhile the plant strives to replace the old withered leaves by developing fresh ones inside the rosette. During this process the neck of the root is elongated and becomes conical, while the lower part grows but little. The disease can also occur on the bloom, if it develops at all, and on the seed-clusters.

When the spots attain a grey colour, there may generally be found in them bunches of fungoid threads belonging to *Cercospora beticola*. From the points of the threads are set free long, narrow, many-jointed conidia that readily germinate and diffuse the disease.

The disease attacks sugar-, fodder-, and red-beets. Amongst the last-mentioned, several sorts have more resistance than others.

PROTECTIVE MEASURES.—(1) Sprinkle those parts of the field that show signs of the disease with Bordeaux mixture (1 per cent.). (2) Do not use diseased beets for seed-beets.

Similar leaf-spots are caused by **C. Apii** on carrot, parsnip, and others, and by **C. radiata** on lady's-fingers (*Anthyllis Vulneraria*).

(ε) RHIZOCTONIEÆ.

Fungi belonging to this group form a felt-like, violet or brown mycelium on the surface of roots and other underground parts. The organs of propagation are poorly developed.

Root Felt Disease. (*Rhizoctonia violacea*.)

This disease—also called " Copper Web " and " Root-Rot "—attacks a great number of plants, especially carrot, beet, clover, and lucerne ; but sometimes also turnip, rape, spurry, and others, and may extend to the roots of trees and shrubs.

In the month of June or July the disease appears on clover and lucerne. There appear circular patches over the fields with yellow and withered plants. On the leaves and stalks nothing is to be found that would give a clue to the phenomenon. But if the plants be pulled up by the roots, there is found on these a red felt of fungoid threads.

On the carrot- and beet-fields the disease becomes conspicuous later on at crop-time. In sundry places the leaves wither away. In this case also the root is covered by a red fungoid felt. This forms a homogeneous layer round either the upper, middle, or lower part of the root. Those parts of the root that are only slightly or not at all affected develop fairly well and retain their natural colour. The felty parts are often shrunk and narrower than the sound parts.

When pulling up the plant, the soil sticks to the fungoid felt, and upon attempting to free the root a part of the fungoid felt falls off as well as some of the root-tissue.

When advanced in age, the threads of the fungoid felt contain a red dye-stuff, but the walls are usually colourless. In the felt are embedded numerous round, brownish-black dots, similar in appearance to the pycnidia of other fungi. These formations never develop any organs of propagation on the living plant. But the following spring, on the dead root, there may be detected in the dark red fungoid meshes numerous egg-shaped conidia with red contents.

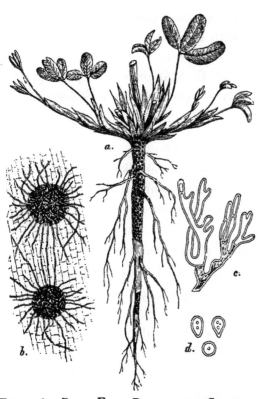

FIG. 106.—ROOT FELT DISEASE OF CLOVER—
Rhizoctonia violacea. (FROM E. ROSTRUP.)

a, Diseased clover-plant; *b,* mycelium and pycnidial knots of threads; *c,* part of a mycelium (from the root of a species of Rumex); *d,* conidia.

Sometimes ramifications of filamentous tubes twine together into irregular bunches, reddish-yellow on the outside, colourless in the middle, and ultimately blackish-red—altogether a kind of *sclerotia.* These formations often separate from the root, mix with the surrounding soil, and become factors in the spread of the disease. From their surface are often separated spool-formed, colourless conidia, either unicellular or bicellular. The mycelium remains alive in the earth from year to year.

It appears as if the various forms of this fungus, which attack diverse host plants, are biologically different. The form that troubles carrot can only with difficulty be conveyed to leguminous

FIG. 107.—ROOT *FELT* DISEASE. (THE AUTHOR.)

a, On carrot ; *b*, on sugar-beet (infected by contagion from a carrot) ; *c*, on fodder-beet ; *d*, on turnip.

plants, beet, and potatoes; at any rate, it will not remain long on these plants.

Efforts have been made to prove a continuous stage of development of this fungus, but the result arrived at has been neither certain nor unanimous. At one time it was surmised to belong to *Trichosphæria circinans* (*Trematosphæria c.*, *Leptosphæria c.*), at another time to *Corticium vagum*, and again to a species of the genus *Rosellinia*.

PROTECTIVE MEASURES.—(1) At harvest-time carefully separate all roots that show even the slightest sign of disease. (2) During the winter examine repeatedly stored-up roots that have been taken from diseased fields, and destroy everything that is unsound. (3) Do not cultivate in diseased soil the same sort of plants for at least three to four years. (4) Should the disease turn up in the first year's clover or lucerne, then the diseased parts of the field should be dug up and sown with grass-seed, and the extension of the disease should be checked by digging trenches round the affected parts of the field. (5) Choose for the cultivation of different plants such sorts as locally have proved of greatest resistance. (6) Take care that the soil is nutritious and well drained.. (7) Do not fertilize with fresh barn manure if the stock has been fed with partly diseased roots.

The related fungoid form **Rhizoctonia Solani** yields small, dark brown papillæ on the surface of potato-tubers. These papillæ consist of a compact mass of reddish-brown, thick, jointed fungoid threads, without any sign of propagating organs. As a rule the fungus does not penetrate deep into the potato, and the papillæ can easily be scraped off.

FIG. 108. — RHIZOCTONIA SOLANI. (FROM E. ROSTRUP.)

Occasionally it may gain an entrance through insect burrows, and then develop a sort of *sclerotia*.

R. fusca forms on the roots of rape and turnip brown, and ultimately almost black, concave papillæ, often uniting into crests. The papillæ consist of short-jointed, ramified fungoid threads, which are entangled at the joints. This fungus remains only on the surface of the root.

CHAPTER XVII

UNEXPLORED DISEASES

Heart Rot of Beet.

THIS disease—also called " Blight of Beet," " Herzfäule," " Maladie du Cœur," " Pourriture du Cœur "—appears in the month of July

a   b

FIG. 109.—HEART ROT OF SUGAR-BEET.   (THE AUTHOR.)

a, Earlier stage of the disease (in the month of August); *b*, later stage of the disease (in the month of October).

or August, on sundry plants, or on certain parts in the beet-fields. The first indication is that the youngest leaves in the rosette turn black and die. Soon it extends to neighbouring older leaves through their petioles, where often broad, pale cross-stripes appear, and so the disease reaches the blades. Gradually these leaves die off, and the beets by the end of the summer have lost all their original leaves. In their place small leaf-rosettes with stunted leaves frequently develop on the neck of the root. At the same time the disease becomes conspicuous on the root, usually first on the outside of the thickest part. It forms there brown, decaying spots that go more or less deeply into the root.

Sometimes the disease stops of its own accord, but the beet becomes inferior to sound specimens both with regard to size and sugar - qualities. Generally the beet decays.

This disease attacks both sugar- and fodder-beets, and varies in its severity in different years, even in the same locality. It is generally

Fig. 110.—Heart Rot of Fodder-Beet. (The Author.)

supposed that prolonged drought during the summer, when the leaf grows most abundantly, renders the plants especially susceptible to the disease, an assisting cause also being that the evaporation from the leaves is too great in comparison with the quantity of moisture introduced by the roots.

As yet nothing is known with certainty about the original cause of this disease. Some think that it is simply a state of general debility in the beet-plant, brought about by excessive cultivation, and that consequently the fungoid formations that are found on the affected beets are only of secondary importance in the work of destruction.   But as a rule fungi of one form or another are considered as being the primary cause.   Several investigators think it to be *Phoma Betæ* (*Ph. sphærosperma, Phyllosticta tabifica*), a stage of development of the genus *Mycosphærella.*   Other investigators seek to find it in *Sporidesmium putrefaciens,* a form of development of the genus *Pleospora,* and still others in a bacterium, *Bacillus mycoides.*

Finally, the opinion has been expressed that the cause of the disease may be a slime mould-fungus, *Myxomonas Betæ,* which at

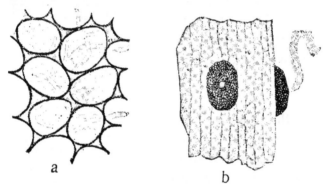

a   b

FIG. III.—PHOMA BETÆ.   (FROM G. LINHART.)

a, Cell-tissue with fungoid mycelium · *b,* pycnidia, one of them extending strings of conidia.

first could in the stage of plasma dwell symbiotically with the plasma of the cell—hence something like the previously described mycoplasm, that is, a latent stage of development of certain rust-fungi—and later on develop as a parasite.   However, the existence of such a mucous fungus has been disputed.

It is possible that the name " Heart Rot " may embrace several separate diseases, each one caused by different agents.

The disease can with certainty be traced back to about 1885, when it appeared very malignantly on beet-fields in Brandenburg and Silesia.   In the year 1892 it was noticed and described in France, and the year after it appeared very destructively in Germany, in almost every district where beets were cultivated. Simultaneously it was noticed also in Belgium. In 1907 it appeared in the southern part of Sweden. Now it is scattered all over Europe.

PROTECTIVE MEASURES.—(1) The earth should be worked deeply, so as to be able to resist the drought. (2) Avoid fertilizers that bring on a too rapid maturing of the beets. (3) Do not sow too early, lest the summer drought affect the beets during their most critical period. (4) Do not throw leaves of diseased beets amongst the manure. (5) For at least four years no beets should be cultivated in a field that has yielded a diseased crop. (6) Only seed-beets should be used that have proved of greatest resistance.

Leaf-Roll Disease of Potato.

Towards the end of July or beginning of August there may be seen on the potato-fields many plants of a peculiar appearance. The leaflets folding over and the edges turning upwards assume a trumpet-like form. Sometimes these plants are scattered over the field, or they are gathered together, forming larger or smaller patches. As the under side of the leaves is turned outwards, its paler colour becomes conspicuous against the dark green hue of the sound plants, and shortly the whole field has a spotted appearance. With many sorts of potatoes the rolled-up leaves have a yellowish or reddish tint, especially the nerves on the under side.

In the first year of disease, the affected plants yield an almost normal crop of sound tubers, being only somewhat defective in starch. If these tubers are used for seed next year, then the eyes will sprout, but when appearing above the ground the plants become stunted, and the rolling up of the leaves commences earlier than in the first year. Tubers are formed, but they are small, often not larger than hazel-nuts. Sometimes the runners get so short that the tubers are close against the stalk, and none of them attain a normal length. The eyes of these small tubers may send out runners, which form rudimentary tubers. If tubers from the second year of disease are used for seed the following year, then their runners cannot reach above the ground, and they may not have any at all.

In the cross-section of the stalk of a diseased potato-plant it will be seen that the vessel-ring is occasionally yellow-tinted.   On a thin section of such a stalk may be seen in the yellow ring a fungoid mycelium that upon cultivation brings forth spores, which might

a

b

Fig. 112.—Leaf-Roll Disease of Potato. (From R. Schander.)
a, Diseased plant ; *b*, crop of Magnum Bonum from sound plants (on the left) and from diseased plants (on the right).

be classified as belonging to the fungus group *Fusarium*.   A similar yellowish ring of vessels is occasionally found inside the diseased tuber, especially at that point where the tuber was fixed to the runner.   But no mycelium has, as a rule, been found, either in the vessel-ring of the tuber or in any other of its tissues, a phenomenon which has been explained by the suggestion that the fungus possibly may be only plasma in the tubers, but develops fungoid threads in the runners.

In many places doubt has been expressed about the parasitical nature of the disease, as there is no mycelium in the diseased tuber. It is supposed that the disease arises from abnormal and unfavourable conditions of the atmosphere and the soil, in connection with the use of immature seed-potatoes. As a result of these co-operating causes, there may have taken place inside the potato-tuber enzymotical disturbances, hence a disorder which possibly is nothing but a recurrence of the long-known " Curl Disease," a malady proved to be only of a temporary nature. In those cases where a fungoid mycelium has been discovered, this should by no means be considered as the origin of the disease, but as something of later growth.

But against this hypothesis that the disease is non-parasitical is the experience of the autumn of 1908. It then appeared in numerous thoroughly investigated localities in Germany, Austria-Hungary, Switzerland, Holland, and other countries, and the outbreak was not due either to climatic or geological causes, and was not due to the use of immature seed-potatoes. As for the theory that the disease is parasitical, it may be observed that the plants which grow from diseased potatoes almost without exception develop the disease, and this occurs even when they have been raised from seed taken from such plants and sterilized; also there is the fact that the disease can be carried to a previously sound field by means of diseased seed-potatoes.

This disease was first noticed in Western Germany, Westphalia, and the Rhine Provinces in 1905, and it has been suggested that this was the effect of an unusually severe and prolonged drought during 1904 and 1905. It soon appeared in other places, and is now known in most European countries.

At the beginning of 1908 this disease induced a genuine panic in Germany and the adjacent countries. This was caused by an alarming newspaper article, written by one of Germany's most prominent potato cultivators, which was reproduced by numerous papers, and headed "The Potato-Culture of Europe at Stake." It expressed the fear that in all Germany—with very minor exceptions —there was not to be found one sound potato for seed. The loss for Germany alone was estimated at 600,000 German marks, or £30,000, and it was considered desirable that the Government should grant 250,000 marks, or £12,500, for the purpose of an experimental station of 500 to 600 hectares, or 125 to 150 acres. These apprehensions proved to be exaggerated. To be sure, the disease appeared in the autumn of 1908 in numerous places in all those European countries where potatoes are cultivated, especially where the industry is carried on extensively, and where the disease once gained a footing there it recurred and attacked all sorts of potatoes. But there were numerous and large districts where it scarcely put in an appearance, and where the crop was an average one.

In the year 1909 the disease appeared in many places in Europe, especially in the south-eastern part of Germany and in Austria-Hungary, and caused greater devastation than in the previous year. For instance, in Bulgaria it spread to such an extent that not a single district was free. And there it happened that different kinds of potato were differently affected, those imported suffering worse than the indigenous. The destruction varied from 10 to 100 per cent. But the very same year several districts in Germany were comparatively free from this disease.

This disease has been thoroughly investigated in recent years in Germany and Austria, but as yet the real cause cannot be ascertained. It may be that under the name of "Leaf-Roll Disease" are included several different forms of disease, varying both with regard to their economic importance and to their appearance in the development of the leaves, their rolling up, the attachment of the tubers, either with or without stolons, and so forth.

PROTECTIVE MEASURES.— (1) Use perfectly sound seed-potatoes taken from a field where no disease has occurred. (2) Do not cultivate potatoes on a diseased field for two to three years. (3) If the disease breaks out in a field, then all affected plants should be pulled up and destroyed; but if you cannot afford to do this, then they should be labelled, so that their tubers can be separated at the harvest from those taken from sound plants. (4) Take care that no earth from a diseased field is brought by means of people, animals, or utensils to a field intended for next year's potato plantation.

A similar, also unexplored, potato disease is reported from Ireland under the name of **"Yellowing,"** or **"Yellow Blight."** By the middle of July the plants begin to wither, turn yellow, and die prematurely, and the crop is next to nothing. No parasite has been detected. Good drainage and ordinary care will arrest the disease.

Mosaic Disease of Tobacco.

During the summer, two to three weeks after the tobacco-plants have been placed out of doors, many become conspicuous by an abnormal appearance. On the leaves are irregular spots of various shades, some of them dark green and not at all transparent, while others are lighter and transparent. When the leaf is held toward the light the formations have a mosaic appearance. The dark green spots grow rapidly and develop into slightly convex excrescences, while the paler spots remain stationary, and the affected plants become stunted in their growth.

Sound and diseased plants are mixed together without any order. The disease does not of itself extend from plant to plant, but if a piece of a diseased leaf, or sap pressed from such a leaf, comes into contact with a wound on a leaf or stem of a sound plant, then there will develop within three to six weeks (all depending upon the age of the sound plant) a similar disease, although not in the infected organ itself, but on the tender leaves which gradually grow out at the terminals of stem and branches. The poison seems to follow with the circulating sap up to the youngest elements of the tissue, where it first begins its work of destruction. A contamination of this kind can readily take place from plant to plant, when the tobacco-plants are topped in order to check the fructifica. tion. Then the infection takes place by means of the workman's hands or implements. As a rule the side branches of the topped plants get the mosaic disease. The diseased leaves fetch less in the market than sound ones. They cannot be used as covering leaves for cigars, and when used for pipe-tobacco are said to have a strong, unpleasant odour.

The disease. attacks all sorts of Virginia tobacco (*Nicotiana Tabacum*), but does not appear on English tobacco (*N. rustica*).

As yet nothing definite is known about the real cause of this disease. Several investigators think that it is caused by bacteria so extremely minute that neither microscope nor culture renders them visible. Other scientists presume the existence of a contagious fiuid ("contagium vivum fluidum "), which is intimately united with the plasmic cell contents, hence a kind of mycoplasm. Others, again, deny the existence of any· parasitic origin of the disease, which they surmise is brought about by certain dis-

FIG. 113. — MOSAIC DISEASE OF TOBACCO. (FROM D IWANOWSKI.)

turbances in the ordinary nutrition of the plant. There can be either an over-production of enzymotically originated oxidases and peroxidases, or otherwise a production of toxin caused by some external irritation. These disturbances may arise from high temperature, abundant watering, damp atmosphere, poor nutrition, or injudicious selection of seed-plants, and so forth.

This disease has long been known in many countries where tobacco is cultivated, but has not been scientifically studied until about the year 1885 in Holland. Since that time it has been strictly investigated, not only in Holland, but also in France, Russia, North America, and elsewhere. On the tobacco plantations round Stockholm it has been very serious for many years past.

PROTECTIVE MEASURES.—(1) Select only sound plants as seed-producers. (2) Keep capsules and seeds in a suitable and dry place, and avoid their being in the neighbourhood of remains of diseased plants. (3) Get sound soil for the hotbeds where the plants are nursed. (4) Drain the field well. (5) Do not use fresh barn-yard manure, but fertilize preferably with kainite and Thomas-phosphate. (6) Top first all plants that suffer with the mosaic disease, then cleanse and disinfect hands and implements before beginning to top the sound plants; this should be done later on.

It often happens that in the same tobacco plantation where mosaic disease exists there appears another spotted disease, sometimes attacking the very same plants that have already been troubled by that complaint. It is called in Germany "Pocken-krankheit." It appears as numerous, often very tiny, brown or white dried-up spots, scattered over the whole blade. Some investigators consider it to be a special disease, others only a sequence to the mosaic disease.

CHAPTER XVIII

GENERAL PROTECTIVE MEASURES AGAINST THE DISEASES

THE measures that may be employed to combat plant diseases caused by parasitical fungi are principally of two kinds : either *preventive* (prophylactic), or *curative* (therapeutic).

With regard to the growing crops of our cornfields, fodder-grounds, and root-crop fields, their close association makes curative measures both tedious and futile; hence in their case the preventive method is the best. The matter is different with garden plants, as each individual tree or shrub can be given curative treatment.

I.—PREVENTIVE MEASURES.

1. *Sound Seed, taken from Sound Plants.*—In certain cases—as, for instance, ergot and brand in seed of cereals and grasses, sclerotia in the seed of clover, beet, turnip, and mustard, and so forth—an ocular examination will be sufficient. The planter himself can undertake this investigation, or he can send samples to the nearest station for seed-control, and obtain a verdict.

But in many cases a botanical examination will not suffice. The most dangerous diseases may evade microscopical analysis, and although neither spores nor mycelia have been detected, experience has proved that the disease has followed with the sowing-seed. This takes place, for instance, with sowing-seed which carries the disease of loose smut and rust, especially yellow rust, and also with potatoes from plants suffering with bad leaf-roll disease. To

178

be perfectly protected against these diseases, one should be certain that the crop from which the sowing-seed is taken was sound, and in some cases it is not enough to know only about the previous year, as the disease might be latent in the sort for several years until a year occurs with atmospheric conditions that provoke an outbreak. Proofs of this have been obtained through several kinds of wheat which are especially susceptible to yellow rust. The sowing-seed of such wheat, being either shrunk or full-sized, might yield apparently a sound crop of good-looking, well-developed seed-corns, and might do so several years successively. But if there comes a year with atmospheric conditions suitable for the development of the fungus of yellow rust, then the disease will break out again in a malignant form. In such cases it will be best for the planter himself to examine the crops of preceding years. If this cannot be done, he should procure reliable information concerning previous crops.

2. *Selection of such Species of Agricultural Plants as possess Resistance.*—It will frequently be found that different varieties of the same sort of plants vary with regard to their susceptibility to the diseases. This takes place, for instance, with the potatoes against the leaf-mould, and wheat against the yellow rust. The best guide in this respect will be the previous experience gained in the place or its vicinity. In different districts the same sort of plants may be susceptible in a different way.

But in other cases it might happen that all the varieties get affected, as is the case with all plants with regard to bacteriosis, and cereals as regards black and brown rust, and so forth.

3. *Careful Preparation and Draining of the Field.*—It goes without saying that too much moisture as well as too much drought render the plants more susceptible to diseases. Hence every precaution should be taken to avoid everything that might deprive them of their stamina.

4. *Fresh Barn-yard Manure should not be used*, as it easily might carry infection.

5. *Diseased Sprouts, Leaves, Roots, and so forth, should be destroyed,*

either by burning or being dug down deeply, otherwise they might diffuse the infection.

6. *Frequent Inspection of Roots stored up for the Winter.*—Everything found to be diseased should be totally destroyed, lest it might in one way or another become a means of diffusion of disease during following seasons.

7. *The Parasites might be Starved.* — Through letting a longer period pass before the same sort of plants again are cultivated on the same place, the parasitical fungi will be rendered destitute of nourishment. This is especially effective with root-parasites such as clump-root on cabbage-plants, root-felt disease on carrot, beet, clover, etc.

8. *All Plants carrying the Infection with them should be removed from the Vicinity.*—This is especially important with regard to cereals and grasses that may be attacked by the different varieties of cluster-cups which shift their host plants. Such plants as *Berberis*, *Rhamnus*, and *Anchusa* ought not to be allowed nearer than 25 to 50 metres, or 80 to 160 feet, to fields and pastures. Likewise weeds like *Triticum repens, Dactylis glomerata,* and others that bear rust fungi which infect cereals and fodder grasses should be removed.

9. *Treating the Sowing-Seed with Fungicides.*— This can be done in various ways :

(*a*) Steeping with copper sulphate solution; already described in this book (pp. 24, 34, 47).

(*b*) Crystallisation with Bordeaux mixture ; already described in this book (p. 48).

(*c*) Steeping with formalin solution ; already described in this book (pp. 13, 48).

(*d*) Ceres treatment ; already described in this book (p. 49).

(*e*) Steeping with sublimate solution ; already described in this book (pp. 13, 124).

(*f*) Warm-water treatment ; already described in this book (pp. 49, 53, 107). For this purpose there are available several kinds of apparatus.

One of these is the " Getreide-Beizapparat," by Appel and

Gassner, manufactured by Paul Altmann in Berlin (Luisen-Strasse 47). This one consists of two principal parts: the warm-water cistern and the steeping cylinder. An ordinary wooden vessel in good condition might be used as a warm-water cistern to hold at least twice as much as the steeping cylinder—say 200 litres, or about 40 gallons.

a   b

FIG. 114.—APPEL AND GASSNER'S APPARATUS FOR STEEPING THE SOWING-SEED. (FROM O. APPEL.)

a, Steeping cylinder with the warm-water cistern above; *b*, letting out the steeped seed.

The steeping apparatus consists of a framework on wheels, and on this the cylinder is hung. Inside this cylinder there are two strainers, one at the bottom and the other 20 centimetres, or 8 inches, below the lid, and is fixed to and removed with it. The space between these two strainers is intended for the seed that is to be treated. The cylinder is supplied with two pipes: one, carrying the water in, enters the cylinder beneath the lower strainer; the other one, leading the water away, extends from the cylinder above the upper strainer.

The warm-water cistern is placed high up, about 4 metres, or 13 feet, on any sort of landing, and is connected by means of a tube with the feeding-pipe of the cylinder. The cistern is filled with water that is heated by steam to a temperature of 55° C. If no steam be available, cold and hot water must be mixed in the cistern until the desired temperature is reached.

When using the apparatus it should be done in this way: The seed is poured into the cylinder above the lower strainer. Then the upper strainer is fixed to the lid, which is put on. The tap is turned, letting in the warm water. This pours into the cylinder from beneath, and runs out through the waste-pipe at the top. At first it will be found that the temperature of the waste water is considerably lower than it was in the feed-pipe. But it will be found that the temperature of 55° C. is attained by the waste water in two minutes after a volume of water one and a half times as large as the quantity of seed has been used. The feeding-tap is then closed, and the apparatus is let alone about five to ten minutes. To counteract any injurious after-effect of the heating up, cold water is let in the cylinder by means of the same tube. After a few minutes, and as soon as the cylinder water is found to be of the same temperature as the cold water in the feeding-pipe, the tap is again closed and the water allowed to run off. Finally, the lid, together with the top strainer, is lifted up, and the seed is poured from the cylinder, and spread out to dry.

The whole process requires only twelve to fifteen minutes, hence there will be time to do four different turns in a single hour. Should the cylinder hold 115 litres, or 23 gallons, of seed, then 50 hectolitres, or 135 bushels, can be treated in one day.

II.—Curative Measures.

When a disease is to be stamped out over large fields or pastures, it is out of the question to pay special attention to each individual plant, as can be done with shrubs and trees in a garden. The only practical method is to sprinkle the whole field with some fungicide.

(a) **Powders as Fungicides.**

10. *Flowers of Sulphur.*—This should be strewn over the diseased plants when the weather is clear, calm, and warm—not colder than 20° C. Then the acid will develop from the sulphur, and kill exterior mycelia—for instance, that of the mildew fungi. The sulphur should be pure and finely ground, then the powder will stick better to the plant. The chemical decomposition is thus promoted and the destruction of the fungi more certain. This method was greatly employed about 1850 in the vineyards of Southern Europe for the destruction of the vine mildew. In recent times this remedy has been superseded by fluids, especially the Bordeaux mixture. This sulphur treatment is also useful against other forms of mildew, as that on peas, and is still the remedy most in use for rose mildew in hothouses.

11. *Sulfosteatit* is another granulated powder, used in the same way as sulphur. It consists of 10 per cent. copper-vitriol and 90 per cent. magnesia. The copper-vitriol is the killing substance, and the magnesia causes it to stick to the plants. This remedy was introduced into the market in 1890 by the firm Jean Souheur of Antwerp. For the diffusion of fungicide powders there have been constructed several kinds of bellows, some used by hand, others carried on the back, and, again, others conveyed by carriage.

(b) **Liquids as Fungicides.**

About the year 1880 liquid fungicides were to some extent employed in the vineyards of France. Later on they gained a footing in England and North America, and it is mainly through extensive experiments in the latter continent that their value has been recognized and utilized. Particularly potato blight is amenable to this sort of treatment.

The most prominent of these fluids is—

12. *Bordeaux Mixture.*— This preparation, made of sulphate of copper, should not be bought in the form of a powder, but as large blue crystals. For the usual mixture (known as 1 per cent.), 1 kilogramme, or 2¼ pounds, of sulphate of copper should be used. This

quantity is placed in a bag of coarse cloth, and submerged in a vessel
containing 50 litres, or about 10 gallons, of water. This vessel should
be a wooden one, not metal, and for the purpose of stirring up the
mixture either wood or glass should be used. It requires about
twelve to twenty-four hours to dissolve, depending upon the tem-

FIG. 115.—PREPARATION OF BORDEAUX MIXTURE. (FROM B. T. GALLAWAY.)
1, Vessel for the lime-wash ; 2, vessel for the copper-vitriol solution ; 3, vessel
for the mixture.

perature of the water. It dissolves more quickly if the bag be
moved backwards and forwards in the fluid.

At the same time a similar quantity of lime-water is prepared
in another vessel. This is done in the following manner : $2\frac{1}{4}$ pounds,
or 1 kilogramme, of quicklime is first slightly sprinkled with water,
then gradually diluted until 50 litres, or 10 gallons, of water have
been used. It is then strained, and no coarse particles are allowed
to remain in the fluid.

Both these solutions are now mixed in equal proportions. The
liquid thus obtained is called Bordeaux mixture because it was first
used in the vineyards surrounding that town. It should be blue
in colour (not green), turn red litmus-paper blue, and, when left
unstirred in a test-tube, deposit a blue sediment at the bottom.
Above this sediment there should be a clear liquid. If it is bluish,

then more lime should be added. A little too much lime will not hurt; but too much copper-vitriol would be harmful, as the acid might burn spots on the plants.

FIG. 116.—KNAPSACK SPRAYER.
(*Benton and Stone, Birmingham.*)

Before filling the sprayer, the mixture should be stirred up thoroughly, as it is the sediment, and not the liquid, that is effective. Mycelia of fungi are thereby killed and germinating spores are checked. As a result, the spread of the disease is stopped. The spraying should take place when the weather is dry.   Should a heavy shower occur immediately after the spraying, then it must be done over again, as the rain washes away the mixture.

Bordeaux mixture of this sort is said to be of 1 per cent.   Should either weaker or stronger mixture be required, then the ingredients should be increased or decreased in proportion.

Freshly-prepared Bordeaux mixture should be used each time the spraying is done, as it loses power.   Recently, however, experi-

FIG. 117.—SPRAY CART WITH TEN-BRANCHED SPRAYER.

ments have been successfully carried out with a view to preserving it for longer periods.   Sugar has proved to be quite effective.   For 1 hectolitre, or 2¾ bushels, of rather weak mixture, 10 to 20 grammes of sugar has retained its fungus-killing qualities for a whole year. But for stronger mixtures—say 2 and 3 per cent.—30 to 50 grammes of sugar should be used for 1 hectolitre, or 2¾ bushels, of the mixture. The sugar should be added within twenty-four hours.

Amongst other mixtures might be mentioned Burgundy mixture (copper sulphate + sodium carbonate), ammoniacal copper carbonate (copper carbonate + ammonia), liver of sulphur (potassium sulphide), and so forth.   But these have mostly been used for garden plants.

APPENDIX

TABLE OF THE *FUNGOID* DISEASES OF AGRICULTURAL PLANTS ARRANGED AFTER THE*IR* HOST PLANTS

I. CEREALS AND GRASSES.

Wheat, *Triticum vulgare.*

A. ON SPROUTS.   PAGE

(*a*) Web-like mesh over the plants in the spring, when the snow melts : **Snow Mould,** *Nectria graminicola* (*Fusarium nivale*) - - - - - - - - - 123

(*b*) Small and hard sclerotia of a reddish-yellow colour on the leaves, while the snow melts : *Typhula graminum* - - 90

B. ON EARS.

(*a*) Corns filled with a brownish-black fetid mass : **Stinking Smut,** *Tilletia caries* and *T. levis* - - - - - 45

(*b*) Spiculæ transformed into a black dust, soon scattered by the wind : **Loose Smut,** *Ustilago Tritici* - - - - 50

(*c*) Ears partly, with an entanglement in the middle, empty and black : *Dilophia graminis* - - - - - 112

(*d*) Seeds of a rosy hue, shrivelled and frequently hollow : **Bacteriosis,** *Micrococcus Tritici* - - - - - 17

(*e*) Awns covered with brick-coloured mucous blotches : **Mucous Mould,** *Fusarium avenaceum* - - - - 153

(*f*) Awns with brown, black-dotted spots : *Phoma Hennebergii* - 148

C. ON LEAVES AND STALKS.

(*a*) Pale spots, with small, black pricks, often arranged in rows : **Leaf-Spot,** *Septoria graminum* and *Ascochyta graminicola*
149, 150

(*b*) Black prickles on the sheaths and leaves (most conspicuous if they be held against the light) : **Black Pricks,** *Leptosphæria Tritici* - - - - - - - - 101

187

C. On all Parts above Ground.

Rye-Grass, *Lolium perenne, L. multiflorum, L. temulentum.*

PAGE

(*a*) **Stinking Smut,** *Tilletia Lolii.* See Wheat, B (*a*) - - 50
(*b*) **Mucous Mould,** *Fusarium avenaceum.* See Wheat, B (*e*) - 153
(*c*) Sterile mycelium in the wall of the grain (especially with
    Lolium temulentum): *Stromatinia temulenta (Endoconidium
    temulentum)* - - ·· - - - - 154
(*d*) *Typhula graminum.* See Wheat, A (*b*) - - - 90
(*e*) **Black Rust,** *Puccinia graminis.* See Wheat, D (*a*) - - 68
(*f*) **Crown Rust,** *Puccinia coronifera.* See Oats, C (*b*) - - 75
(*g*) **Mildew,** *Erysiphe graminis.* See Wheat, D (*d*) - - 94
(*h*) **Leaf Spot,** *Septoria graminum.* See Wheat, C (*a*) - - 150
(*i*) Small brown spots on the leaves : *Gloeosporium graminum* - 151
(*j*) **Ergot Disease,** *Claviceps purpurea.* See Rye, B (*b*) - - 126

Meadow Fescue, *Festuca elatior.*

(*a*) **Timothy Rust,** *Puccinia Phlei - pratensis.* See Timothy
    Grass, (*a*) - - - - - - - 83
(*b*) **Crown Rust,** *Puccinia coronifera.* See Oats, C (*b*) - - 75
(*c*) **Ergot Disease,** *Claviceps purpurea.* See Rye, B (*b*)- - 126

Strand Fescue, *Festuca arundinacea.*

Ergot Disease, *Claviceps purpurea.* See Rye, B (*b*) - - 126

Brome Grass, *Bromus arvensis, B. mollis,* and others.

(*a*) **Stinking Smut,** *Tilletia Holci.* See Wheat, B (*a*) - - 50
(*b*) **Loose Smut,** *Ustilago bromivora.* See Wheat, B (*b*) - - 60
(*c*) **Brown Rust,** *Puccinia bromina.* See Wheat, D (*b*) - - 84
(*d*) **Mildew,** *Erysiphe graminis.* See Wheat, D (*d*) - - 94

Meadow-Grass, *Poa pratensis, P. compressa, P. trivialis,*
P. nemoralis.

(*a*) Leaves with long, breaking wound-stripes, filled with a brown-
    ish-black dust-mass : **Smut,** *Tilletia striæformis* - - 50
(*b*) **Black Rust,** *Puccinia graminis.* See Wheat, D (*a*) - - 68
(*c*) Small, yellow, scattered, dust-filled sores, mostly on the upper
    side of the leaves ; finally small, black dots, covered by the
    epidermis of the leaf, especially on its lower side : *Puccinia
    Poarum* and *Uromyces Poæ* - - · - 78, 86
(*d*) **Mildew,** *Erysiphe graminis.* See Wheat, D (*d*) - - 94
(*e*) **Reed Mace,** *Epichloë typhina.* See Timothy Grass, (*b*) - 125

Reed Poa Grass, *Glyceria aquatica.*

False Oat Grass, *Avena elatior.*

Yellow Oat Grass, *Trisetum flavescens.*

Bent Grass, *Agrostis vulgaris,* Fiorin Grass, *A. stolonifera,* and others.

Smallbreed Grass, *Calamagrostis arundinacea, C. lanceolata,* and others.

Spreading Milium Grass, *Milium effusum.*

Soft Grass, *Holcus mollis, H. lanatus.*

II. ROOTS.

Potato, *Solanum tuberosum.*

A. ON TUBER AND ROOT.

B. ON STEM AND LEAVES.

(*c*) Leaves—especially while young—thick, curled, often with rolled-up edges ; when older, covered on the under side with a slate-grey mould : **Mould**, *Peronospora Schachtii* - - 39

(*d*) Leaves and stalks covered with a thin white film, with numerous embedded black dots : **Mildew**, *Erysiphe Polygoni* - 95

(*e*) Leaves with large pale yellow, red-brimmed patches, sprinkled with brown prickles : *Phyllosticta Betæ* - . - 144

(*f*) Leaves with brownish-grey, often red-brimmed, spots on the under side, these being covered with fine grey bunches of fungoid threads : *Cercospora beticola* - - - - 162

(*g*) Leaves with round, red-violet spots, which are white in the middle owing to a nap of fungoid threads : *Ramularia Betæ* 159

(*h*) Leaves with brown, dry patches : *Sporidesmium putrefaciens* (*Pleospora putrefaciens*) - - - . - 104

C. On Root.

(*a*) Vessel strings in the root, first reddish-brown, then black ; the interior of the root finally dissolved into a glutinous or molasses-like slime ; top of the root black, dead : **Mucous Bacteriosis**, *Bacillus Betæ* and others - - - 13

(*b*) Small black warts on the surface of the root, often together forming vertical swellings : **Wart Bacteriosis**, *Bacterium scabiegenum* - - - - - - - 15

(*c*) Large, short-stemmed, irregularly intersected, tumour-like excrescences on the upper part of the root : **Beetroot Tumour**, *Urophlyctis leproidea* - - - - 30

(*d*) Wart-like excrescences or fissure-like concavities on the surface of the root ; root usually with an entanglement at the middle : *Oospora cretacea* and others - - - - - 158

(*e*) Decaying spots, with white, afterwards yellow, slimy, shrivelled covering : *Fusarium Betæ* - - - . - 157

(*f*) **Root Felt Disease**, *Rhizoctonia violacea*. See Potato, A (*e*) 163

D. On the Whole Plant.

(*a*) **Root-Crop Rot**, *Sclerotinia Libertiana*. See Potato, B (*b*) - 135

(*b*) Leaves small, yellowish-brown, finally black ; root dwarfed, with dark concentric rings in the interior ; from the vessels a suppurating dark juice : **Californian Beet-Pest** - - 16

(*c*) Small, brown sclerotia on the upper part of the root, and also on and within the stem and branches : *Typhula Betæ* - 90

(*d*) The youngest leaves in the rosette first brown, then black ; root ceasing to grow, finally its surface is covered with con-

[1] Here are included—in order to avoid repetition—the roots Swedish .Turnip and Turnip, as well as Cabbage and Rape, the injurious fungi of all cruciferous plants being essentially the same.

C. On Parts above Ground.

(*a*) Leaves and stem with small, rusty-brown, finally black, dust-filled sores : **Rust,** *Uromyces Pisi-sativi* - - - 87

(*b*) Leaves with large, discoloured patches, covered on the under side with a thick grey-violet blight : **Blight,** *Peronospora Viciæ* - - - - - - - 40

(*c*) **Mildew,** *Erysiphe Polygoni.* See Beet, B (*d*) - - 95

(*d*) **Stem Mould,** *Sclerotinia Fuckeliana.* See Potato, B (*a*) - 133

(*e*) Stalks and fruits, at times also seeds, covered with yellow, brown-edged spots : *Ascochyta Pisi* - - - - 149

(*f*) The whole plant suddenly turning yellow, covered with a greenish-black dust : *Cladosporium herbarum* - - 115

Vetch, *Vicia sativa, V. villosa,* and others, **Windsor-Bean,** *Faba vulgaris.*

A. On Root.

Vessel Brand, *Fusarium vasinfectum.* See Common Pea, B (*b*) 155

B. On Parts above Ground.

(*a*) Leaves and stem with rusty-brown, dust-filled sores ; later on with hard, blackish-brown swellings : **Rust,** *Uromyces Fabæ* 88

(*b*) **Blight,** *Peronospora Viciæ.* See Common Pea, C (*b*) - 40

(*c*) **Mildew,** *Erysiphe Polygoni.* See Beet, B (*d*) - - 95

(*d*) **Stem Mould,** *Sclerotinia Fuckeliana.* See Potato, B (*a*) - 133

(*e*) *Ascochyta Pisi.* See Common Pea, C (*e*) - - - 149

(*f*) Pale yellow patches on the leaves, sprinkled with brownish-black dots : *Phyllosticta Fabæ* - - - - 144

(*g*) Leaves with round, white spots, surrounded by a brown border ; on the under side of the leaves a nap of fungoid threads : *Ramularia sphæroidea* - - - - 160

Lupin, *Lupinus luteus, L. albus, L. angustifolius.*

A. On Seedlings.

(*a*) Leaves with first yellow then brown spots ; plants withering away : **Bacteriosis,** *Bacillus elegans* - - - 17

(*b*) **Seedling Blight,** *Pythium Baryanum.* See Barley, A - 33

B. On Root.

(*a*) **Root-Rot,** *Thielavia basicola.* See Common Pea, B (*a*) - 97

(*b*) **Vessel Brand,** *Fusarium vasinfectum* [see Common Pea, B (*b*)] and *Fusarium roseum,* var. *Lupini albi* - - 155, 157

(*i*) Leaves with brown spots; upon these spots small reddish-brown waxy discs : **Leaf-Spot,** *Pseudopeziza Trifolii* - 140

(*j*) Leaves with dark brown spots, extending more and more until the whole leaf is dried : *Macrosporium sarcinæforme* - 112

(*k*) Leaves with small, white, black-pricked spots : *Phyllosticta Trifolii* - - - - - - - 144

(*l*) Leaves on the upper side with small, light brown spots, surrounded by a purple-reddish brim : *Sphærulina Trifolii* - 119

Lucerne, *Medicago sativa, M. lupulina,* and others.

A. ON ROOT.

(*a*) **Root Felt Disease,** *Rhizoctonia violacea.* See Potato, A (*e*) 163

(*b*) **Clover-Sclerote,** *Sclerotinia Trifoliorum.* See Clover, B (*b*) - 138

(*c*) *Mitrula Sclerotiorum.* See Clover, B (*c*) - - - 142

(*d*) On the neck of the root numerous large, irregularly-formed, coral-shaped tumours : **Crown Gall,** *Urophlyctis Alfalfæ* - 30

B. ON PARTS ABOVE GROUND.

(*a*) **Rust,** *Uromyces striatus.* See Clover, C (*a*) - - - 87

(*b*) **Blight,** *Peronospora Trifoliorum.* See Clover, C (*b*) - - 41

(*c*) **Mildew,** *Erysiphe Polygoni.* See Beet, B (*d*) - - 95

(*d*) *Typhula Trifolii.* See Clover, C (*e*) - - - - 91

(*e*) **Leaf-Spot,** *Pseudopeziza Trifolii.* See Clover, C (*i*) - - 140

(*f*) Leaves with small, white, brown-edged, black-pricked spots : *Septoria Medicaginis, Phyllosticta Medicaginis,* and *Pleosphærulina Briosiana* - - - - - 120, 144, 150

(*g*) *Ascochyta Pisi.* See Common Pea, C (*e*) - - - 149

Meadow Pea, *Lathyrus pratensis,* **Earthnut Pea,** *L. tuberosus,* and others.

(*a*) **Rust,** *Uromyces Fabæ.* See Vetch, B (*a*) - - - 88

(*b*) **Blight,** *Peronospora Viciæ.* See Common Pea, C (*b*) - 40

(*c*) **Mildew,** *Erysiphe Polygoni.* See Beet, B (*d*) - 95

(*d*) **Stem Mould,** *Sclerotinia Fuckeliana.* See Potato, B (*a*) - 133

(*e*) Black, round, crust-like spots on the leaves : *Diachora Onobrychidis* - - - - - - - 132

(*f*) Leaves with small, pale spots on the under side, covered by a nap of fungoid threads : *Ovularia deusta* - - - 160

(*g*) Leaves with small, pale, black-pricked spots : *Septoria silvestris* 150

Esparcet, *Onobrychis sativa.*

(*a*) Leaves and stalk with rusty-brown, finally black, dust-filled sores : **Rust,** *Uromyces Onobrychidis* - - - 89

Spurry, *Spergula arvensis.*

(a) **Seedling Blight,** *Pythium Baryanum.* See Barley, A - 33
(b) Leaves and stem with reddish-brown, later on black, round, dust-filled sores : **Rust,** *Puccinia Spergulæ* - - - 86
(c) Leaves with pale spots, covered by a fine grey mildew : **Blight,** *Peronospora obovata* - - - - - 41
(d) Leaves and stalks with pale, later on black, spots ; dying : *Sphærella isariphora* and *Septoria Spergulæ* - 119, 150

Buck-Wheat, *Fagopyrum esculentum.*

(a) **Stem Mould,** *Sclerotinia Fuckeliana.* See Potato, B (a) - 133
(b) Leaves and stem with large, round, buff-coloured spots, with darker brim and drab-coloured centre : *Ascochyta Fagopyri* 149
(c) Leaves and stem with pale spots, surrounded by a light red brim : *Phyllosticta Polygonorum* - - - - 144

Tobacco, *Nicotiana Tabacum.*

A. On Seedlings.

(a) Stem part rotting, the putrefaction progressing from below : **Bacteriosis,** *Bacillus amylobacter* [See Potato, A (a)], and *Olpidium Nicotianæ* - - - - - 2, 31
(b) **Root-Rot,** *Thielavia basicola.* See Common Pea, B (a) - 97
(c) Plants, especially seed-leaves, slack, slimy, finally with a black, velvety covering : *Alternaria tenuis* - - - - 112

B. On the Full-grown Plant.

(a) The joints of the stem, with long, groove like, dark spots ; plant dying : **Stem Bacteriosis,** *Bacillus æruginosus* 17
(b) Leaves with small, pale, finally white and desiccative patches : **Leaf Bacteriosis,** *Bacillus maculicola* - - - 17
(c) Leaves and stem with whitish spots ; on these spots later on hard, black sclerotia : *Sclerotinia Nicotianæ* - - 140
(d) **Mildew,** *Erysiphe Polygoni.* See Beet, B (d) - - 95
(e) Leaves with numerous, light, later on white and dried, spots ; occasionally with little black dots at the centre : *Phyllosticta Tabaci* - - - - - - 144
(f) Leaves with brown, dried, irregular spots : *Ascochyta Nicotianæ* - - - - - - - 149
(g) The youngest leaves with irregular spots of various sizes and shades, some of them light and transparent, others dark and not at all transparent ; older leaves battered, frequently distorted : **Mosaic Disease** - - - - 175
(h) Leaves with small, brown, or white, finally dried patches : **Pox Disease** - - - - - - 175

INDEX

Baillière, Tindall and Cox, 8, Henrietta Street, Covent Garden, London.

An Abridged List of Works

PUBLISHED BY

BAILLIÈRE, TINDALL & COX.

AARONS' Gynæcological Therapeutics.
Crown 8vo. Pp. xiv + 178; with 46 *I*llustrations. Price **5s.** net.

AXENFELD'S Bacteriology of the Eye.
Translated by ANGUS MACNAB, F.R.C.S. Royal 8vo. Pp. xv + 410, with 105 *I*llustrations, mostly coloured. Price **21s.** net.

BIANCHI'S Text-book of Psychiatry.
Royal 8vo. Pp. xvi + 904, with 106 *I*llustrations. Price **21s.** net.

BIDWELL'S Handbook of Intestinal Surgery.
Second Edition. Demy 8vo. Pp. xiv + 215, with 120 *I*llustrations. Price **6s.** net.

BRAND & KEITH'S Clinical Memoranda for General Practitioners. Crown 8vo. Pp. x + 208. Price **3s. 6d.** net.

BROWN'S Physiological Principles in Treatment.
Second Edition. Crown 8vo. Pp. viii + 392. Price **5s.** net.

BUCHANAN'S Manual of Anatomy, Systematic and Practical, including Embryology.
In 2 volumes. Price **12s. 6d.** each net. Or complete in 1 volume. Price **21s.** net. Demy 8vo. Pp. 1572, with 631 *I*llustrations, mostly original, and in several colours. (*University Series.*)

CALDWELL'S Military Hygiene.
Second Edition. Demy 8vo. Pp. xiv + 580, with 80 *I*llustrations, mostly original. Price **12s. 6d.** net.

CALKINS' Protozoology.
Royal 8vo. Pp. xii + 350, with 4 Coloured Plates and 125 *I*llustrations. Price **15s.** net.

CAMPBELL'S On Treatment.
Crown 8vo. Pp. viii + 422. Price **5s.** net.

CASTELLANI & CHALMERS' Manual of Tropical Medicine. Demy 8vo. Pp. xviii + 1242, with 14 Coloured Plates and 373 Illustrations, mostly original. Price **21s.** net. (*University Series.*)

CROOK'S High-Frequency Currents.
Second Edition. Demy 8vo. Pp. x + 206, with 44 *I*llustrations. Price **7s. 6d.** net.

DIEULAFOY'S Text-book of Medicine.

Translated by V. E. COLLINS and J. A. LIEBMANN. In 2 volumes. Royal 8vo. Pp. xxiv + 2082, with 7 Coloured Plates and 105 Illustrations. Price **25s.** net.

ELDER'S Ship-Surgeon's Handbook.

Second Edition. Crown 8vo. Pp. xii + 388. Price **5s.** net.

FRENCH'S Medical Laboratory Methods and Tests.

Second Edition. Pp. viii + 168, with 2 Coloured Plates and 88 original Illustrations. Leather, gilt tops. Price **5s.** net

FREYER'S Clinical Lectures on Surgical Diseases of the Urinary Organs.

Demy 8vo. Pp. viii + 425, with 141 Illustrations, mostly original. Price **12s. 6d.** net.

GARDNER'S Surgical Anæsthesia.

Crown 8vo. Pp. xii + 240, with 12 Plates and 35 other Illustrations. Price **5s.** net.

GEMMELL'S Chemical Notes and Equations.

Second Edition. Crown 8vo. Pp. xiv + 266 Price **5s.** net.

GRAY'S Diseases of the Ear.

Demy 8vo. Pp. xii + 388, with 53 Plates, of which 37 are Stereoscopic, and 70 other Illustrations. Price, with Stereoscope, **12s. 6d.** net.

GREEN'S Pathology.

Eleventh Edition. Demy 8vo. Pp. viii + 642, with many new Illustrations, plain and coloured. Price **15s.** net. (*University Series.*)

HYDE'S Diseases of the Skin.

Eighth Edition. Royal 8vo. Pp. xxviii + 1126, with 58 Plates and 223 Illustrations, Plain and Coloured Price **25s.** net.

JELLETT'S Manual of Midwifery.

Second Edition. Demy 8vo. Pp xiv + 1210, with 17 Plates and 557 Illustrations, plain and coloured. Price **21s.** net. (*University Series.*)

KERR'S Operative Midwifery.

Second Edition. Royal 8vo Price **21s.** net.   [*In the press.*

KNOCKER'S Accidents in their Medico-Legal Aspect.

A Practical Guide for the Expert Witness, Solicitor, and Barrister, by leading Medical Authorities. Royal 8vo. Pp. xxviii + 1266. With 206 Illustrations, Plates, and Diagrams. Price **30s.** net.

KNOX'S Military Sanitation and Hygiene.

Pocket size. Limp leather cover, gilt top. Pp. xii + 346, with 21 Illustrations. Price **5s.** net

LAKE'S Handbook of Diseases of the Ear.

Third Edition. Demy 8vo. Pp. xii + 248, with 4 Coloured Plates and 66 Original Illustrations. Price **7s. 6d.** net

LAMB'S Diseases of the Throat, Nose, and Ear.
Crown 8vo. Pp. xvi + 322, with 55 *I*llustrations. Price **7s. 6d**. net.

LAMBKIN'S Syphilis : Its Diagnosis and Treatment.
With Special Section on the Ehrlich-Hata Treatment " 606." Demy 8vo. Pp. viii + 193. Price **5s**. net.

LAVERAN & MESNIL'S Trypanosomes and the Trypano-somiases.
Translated and Edited by DAVID NÁBARRO, M.D. Royal 8vo Pp. 538, with Coloured Plate and 81 *I*llustrations. Price **21s**. net.

MACEWEN'S Surgical Anatomy.
Demy 8vo. Pp. xii + 452, with 61 *I*llustrations, plain and coloured. Price **7s. 6d**. net.

McKAY'S Operations upon the Uterus, Perineum, and Round Ligaments.
Demy 4to. Pp. xvi + 454, with 148 original Plates. Price **21s**. net.

McKISACK'S Dictionary of Medical Diagnosis.
Demy 8vo. Pp. xii + 584, with 77 *I*llustrations. Price **10s. 6d**. net.

MAY & WORTH'S Diseases of the Eye.
Second Edition. Demy 8vo. Pp. viii + 400, with 336 *I*llustrations, including 22 Coloured Plates. Price **10s. 6d** net.

MONRO'S Manual of Medicine.
Third Edition. Demy 8vo. Pp. xxii + 1024, with 42 *I*llustrations, plain and coloured. Price **15s**. net. (*University Series.*)

MUMMERY'S After-Treatment of Operations.
Third Edition. Crown 8vo. Pp. x + 252, with 38 Illustrations. Price **5s**. net.

ONODI'S Optic Nerve and Accessory Sinuses of the Nose.
Translated by J. LÜCKHOFF, M.D. Small 4to. Pp. viii + 102, with 50 original Plates and *I*llustrations. Price **10s. 6d**. net.

OSTERTAG'S Handbook of Meat Inspection.
Third Edition. Royal 8vo. Pp. xxxvi + 886, with Coloured Plate and 260 *I*llustrations. Price **31s. 6d**. net.

POLITZER'S Text-book of Diseases of the Ear.
Fifth Edition, entirely rewritten. Translated by M. J. BALLIN, M.D., and P. H. HELLER, M.D. Royal 8vo. Pp. xiv + 892, with 337 original *I*llustrations. Price **25s**. net.

RAMSAY'S Diathesis and Ocular Diseases.
Crown 8vo. Pp. viii + 184, with 17 Plates. Price **3s. 6d**. net.

ROBERTSON'S Meat and Food Inspection.
Demy 8vo. Pp. x + 372, with 40 *I*llustrations. Price **10s. 6d**. net.

ROSE & CARLESS' Manual of Surgery.
Eighth Edition. Demy 8vo. Pp. xii + 1406. With 12 Coloured Plates and many new and original *I*llustrations. Price **21s.** net. (*University Series.*)

SCALES' Practical Microscopy.
Second Edition. Crown 8vo. Pp. xvi + 334, with 122 *I*llustrations. Price **5s.** net.

STARR'S Organic and Functional Nervous Diseases.
Third Edition. Royal 8vo. Pp. 912, with 29 Plates, plain and coloured, and 300 Figures in the text. Price **25s.** net.

STEVEN'S Medical Supervision in Schools.
Demy 8vo. Pp. x + 268, with 40 Illustrations. Price **5s.** net.

STEWART'S Manual of Physiology.
Sixth Edition. Demy 8vo. Pp. xx + 1064, with 2 Coloured Plates and 449 *I*llustrations. Price **18s.** net (*University Series.*)

TREDGOLD'S Mental Deficiency.
Demy 8vo. Pp. xviii + 391, with 67 *I*llustrations and 9 Charts. Price **10s. 6d.** net.

TUCKEY'S Treatment by Hypnotism and Suggestion.
Fifth Edition. Demy 8vo. Pp. xviii + 418. Price **10s. 6d.** net.

TURNER'S Radium : Its Physics and Therapeutics.
Crown 8vo. Pp. x + 88. With 20 Plates and other Illustrations. Price **5s.** net.

WALSH'S The Röntgen Rays in Medical Work.
Fourth Edition. Demy 8vo. Pp. xviii + 434, with 172 Plate and other *I*llustrations. Price **15s.** net.

WALTERS' Open-Air or Sanatorium Treatment of Pulmonary Tuberculosis. Demy 8vo. Pp. xvi + 323, with 29 *I*llustrations. Price **5s.** net.

WHEELER'S Operative Surgery.
Second Edition. Demy 8vo. Pp. xiv + 296, with 157 *I*llustrations. Price **7s. 6d.** net.

WHITLA'S Medicine. In two volumes. Price **25s.** net.

„ **Materia Medica.** Ninth Edition. Price **9s.** net.

„ **Dictionary of Treatment.** Fifth Edition.
[*In preparation.*

WILLIAMS' Minor Maladies and their Treatment.
Second Edition. Crown 8vo. Pp. xii + 404. Price **5s.** net.

YOUNGER'S Insanity in Everyday Practice.
Second Edition. Crown 8vo. Pp. viii + 124. Price **3s. 6d.** net.

LONDON :
BAILLIÈRE, T*I*NDALL & COX, 8, HENRIETTA ST., COVENT GARDEN.

Lightning Source UK Ltd.
Milton Keynes UK
UKHW022138300119

336486UK00009B/674/P